Digital Scientific Communication

Ramón Plo-Alastrué • Isabel Corona
Editors

Digital Scientific Communication

Identity and Visibility in Research Dissemination

palgrave
macmillan

Editors
Ramón Plo-Alastrué
Universidad de Zaragoza
Zaragoza, Spain

Isabel Corona
Universidad de Zaragoza
Zaragoza, Spain

ISBN 978-3-031-38206-2 ISBN 978-3-031-38207-9 (eBook)
https://doi.org/10.1007/978-3-031-38207-9

This Palgrave Macmillan imprint is published by the registered company Springer Nature Switzerland AG.
The registered company address is: Gewerbestrasse 11, 6330 Cham, Switzerland

Prologue: State of the Art of Research Dissemination

Abstract The Foreword looks at how academic publications have been influenced by the rapid expansion of digital means of communication and by the global extension of the participation framework provided by the web. In a research context that is increasingly digital and international, identity and visibility have become a major issue, especially with the proliferation of academic publications, the availability of new communicative environments—websites, blogs, social media—and the radical changes that have involved even the traditional key elements of publishing—abstracts and journal articles. Attention is paid to the linguistic dimension of this impact, ranging from issues of communicative inequality to the need to compete for visibility and self-promotion: the new context requires attention to a complex media ecology and to the changes taking place in the whole knowledge system. Another key element of the wider social context that is important to consider is the growing social need for academia to engage with stakeholders and the public. Access to knowledge has become a key issue in a "knowledge society". This involves debates on Public Science—transferring the results of research to professionals or to the wider public, with a view to contributing to society—as well as Open Science—responding to problems of communicative inequality by favouring access to knowledge, for example, with open access publishing and education. The role of the web in engaging with the wider public opens a whole cline of possibilities that further characterise the discourse of knowledge, well beyond the basic distinctions traditionally opposing knowledge communication and popularisation.

Keywords Academic publishing • Digital communication • Media ecology • Access to knowledge • Science communication

The rapid expansion of digital means of communication and the global extension of the participation framework provided by the web have changed academic life and publications profoundly (e.g. Lorés & Diani, 2021). Understanding this change is essential for academics, and particularly so in the world of science. Digital affordances have had an undeniable impact on the ways in which science is communicated, by multiplying the channels and the semiotic modes at the disposal of researchers and readers. In a research context that is increasingly digital and international, identity and visibility have become a major issue, especially with the proliferation of publications that are available. At the same time, as digital communication has drawn attention to issues of access to information, scientists are increasingly asked, on the one hand, to make sure that their research is available to all other researchers and that the presentation of their results is fully transparent and, on the other, they are required to disseminate the content of their research to a wider and often undefined audience. These are major changes in the context of science communication, in which the challenges of self-promotion and knowledge dissemination are found to co-exist in always novel combinations.

Understanding the impact of the digital transformation on academic writing practices requires adopting a wide perspective. It is a perspective that takes all the affordances of digital discourse into consideration—multimodality, hypertextuality, interactivity, anonymity—and that acknowledges the effect of digital media on many old and new genres (Luzón & Pérez-Llantada, 2019) in a complex digital media ecology (Weitkamp et al., 2021). In a wider social perspective, however, it is important to look also at how the whole knowledge system seems to be changing under technological, economic and distributional influences, as well as under the influence of a new emphasis on the social role of knowledge (Cope & Kalantzis, 2014). This wider focus throws light on the questions of identity and visibility in research dissemination that are explored in the volume.

Primarily, digital scientific communication should be understood in the light of the present-day nature of publishing as a massive commercial industry, with the inevitable marketisation of knowledge and the need for self-promotion. Academic publishing has become a massive commercial industry (Hyland, 2016): the 2018 report of the STM (the association of Scientific, Technical and Medical publishers) accounts for over 3 million articles published every year in English-speaking peer-reviewed journals (Johnson et al., 2018). This has not only led to a marketisation of knowledge but also to the emergence of alternative metrics, such as the number of views of an article, its downloads, the presence of a study or a scholar on social media. The expansion of academic publishing in general has also meant an expansion of publishing in English, with all the related problems of communicative inequality (Plo-Alastrué & Pérez-Llantada, 2015), the complex forms of bilingualism required of most researchers, and the increasing number and types of "literacy brokers" (Lillis & Curry, 2010) that play a role in the process of publication. The standards of English for research publication purposes are changing, in the context of international communities where English is mostly a Lingua Franca (Mauranen, 2010). In the framework of a general need to be "always on" (Baron, 2008)—a "fast academia" (Berg & Seeber, 2017) where the omnipresence of work tends to blur the distinction between personal and work life—scholars are pushed towards "continual publishing" (Lockley & Carrigan, 2011) for visibility.

Competition for visibility has brought about changes in the publishing industry and in publishing sites in general, leading to the increasing importance of journal rankings and branding and to the birth of predatory journals (see Bocanegra-Valle, this volume). Even more importantly, the issue of visibility has become central for scholars themselves, often determining a rhetorical foregrounding of novelty and significance in research writing (Hyland, 2016), combined with greater writer assertiveness in many contexts. The "hard sciences," in particular, show an increase in the use of involvement features (interpersonal and evaluative meanings) away from their traditional objective style towards "more involved, stance-laden discourses, which emphasise the role of the interpreting researcher" (Hyland & Jiang, 2019, pp. 227–230).

One of the key elements of the present volume is thus the focus on visibility and self-promotion. The need for self-promotion spans across all sectors (see, for example, Sancho-Guinda, this volume, on technology disclosures). The performance of an academic self—a "persona," a strategically developed public identity—becomes central in the construction of online reputation (Marshall et al., 2017) and in self-branding. This has also increased the attention paid to the visibility potential of a multiplicity of new environments such as blogs (Diani & Freddi, this volume), video abstracts (Dontcheva-Navratilova, this volume) and 3-minute thesis presentations (cf. the three chapters in section D).

Another important issue dealt with in the present volume is the growing social need for academia to engage with stakeholders and the public. Access to knowledge has become a key issue in a "knowledge society." The huge expansion of specialised journals and the speed of dissemination (including pre-prints and forms of self-publishing) has highlighted the need to facilitate public access to research and its data, adopting accessible and transparent processes of knowledge creation and public knowledge dissemination. In this context, two perspectives are inextricably interconnected, usually referred to as "public science" and "open science." In Public Science, scholars are asked to transfer the results of their research to professionals or to the wider public, with a view to contributing to society by shaping communities' development, developing community-based research and aligning curricula with the needs of the community. The international dimension of academic communities has also drawn attention to issues of inequality in access to knowledge and suggested that the results of research should be openly available to everyone.

Open Science is expected to respond to problems of communicative inequality by favouring access to knowledge, for example with open access publishing and education (see Vicente-Sáez & Martínez-Fuentes, 2018 for a literature review). A further development, referred to as "open data," also requires that data are shared with other researchers or citizens, so that research transparency is guaranteed and data are accessible and exploitable by others, thus creating new connections between the traditional research narrative and the evidence, which is no longer simply reported but made accessible.

Open access publishing is now a widely acknowledged practice. Starting with the *Journal of Medical Internet Research* (1999) and especially the *Public Library of Science (PLOS)* (2001), the trend to promote Open Access scientific publishing has acquired increasing centrality over the past 20 years. The pandemic has somewhat intensified the process, highlighting the need for open access of research and data (Lee & Haupt, 2021), as well as the needs and challenges of appropriate public health science (Richardson, 2020).

It should be noted, in fact, that the web makes knowledge potentially available to a large virtual community, but it also creates new problems of information selection for users and clearly shows the limits of purely technical participatory mechanisms. After a long process that over the centuries had separated the expert reader of research publications from the general reader of popular science, the digital media now often reach indeterminate audiences. General readers now have potential access to the same texts as expert readers, but can they cope cognitively? And do they have similar interests? The blurring of the different publication environments does not in itself guarantee cognitive access or equal interests. Changes are brought about in both authorial identity and readership. This often means researchers need to produce a range of texts or textual clusters addressing different audiences and to adopt different textual strategies that may respond to the needs of different readers (see Benelhadj, this volume).

Engagement with stakeholders and the general public determines new environments, new genres and new sensitivity to the needs of public science. On the one hand, for example, there is the development of new communicative formats, such as research websites, social media, or blogs. All of them can help researchers to publicise their research, to engage in networks with other academics, to disseminate information, to increase visibility, to facilitate discussion and to engage with non-academics. Research websites are an important tool in responding to the needs of disseminating research and building researchers' visibility alike, as they allow researchers to construct and display their vision and values through verbal and visual elements, including, for example, evaluative language choice, engagement markers and multimodal layout (Pascual et al., 2020; Lorés, 2020; Corona, 2021; Mur-Dueñas, 2021). Blogs are also interesting tools for visibility and potential areas of dissemination, but they are hybrid contexts in many ways, as they seem to collapse not only spoken and written discourse, elements of monologue and dialogue and different semiotic modes, but they also mix and blur the centre and the periphery of research discourses, as well as the private and the public self of bloggers (e.g. Bondi, 2018a, 2018b, 2018c, 2022). Social media are key elements of public engagement (Jünger & Fahnrich, 2020); Twitter, for example, is thought to facilitate immediate sharing and reach potentially interested audience (Klar et al., 2020), using a range of semiotic resources that facilitate dialogue and the creation of online communities (Zappavigna, 2012; Holmberg & Thelwall, 2014) through affordances such as emojis, mentions (@), hashtags (#) and retweeting.

The impact of digital communication has also produced interesting changes in traditional genres. The central role of abstracts, for example, has actually turned abstracting services into "hubs" that direct researchers to the relevant literature. Journals have exploited all the possibilities offered by a digital environment in terms of intensifying intertextual links and hypertextuality in research articles well beyond citation and other cross-references (Pérez-Llantada, 2016). Journal articles now have audio and video components that either complement the verbal text (as when lengthy descriptions of surgical procedures are substituted by images, video components and animations) or lead to the creation of purely visual and video texts, as is increasingly frequent with abstracts. The new freedom from space restrictions in digital publishing leads to much greater complexity in the structure of an article and allows the creation of Virtual Special Issues, with introductions linking independently produced articles and thus mapping an evolving territory (Mur-Dueñas, 2018).

When looking in particular at the world of research dissemination, the public role of science can be understood within the complex network of Knowledge Communication (Kastberg, 2010), looking at how knowledge is constructed, presented and communicated. The difference between communicating within the community of experts and communicating outside the specific community becomes central (Calsamiglia & van Dijk, 2004) and the empowerment of lay people and non-experts beyond professional circles and academic communities also becomes relevant.

Studies on popularisation have long highlighted the different purposes that guide both writers and readers in public communication, as opposed to communication within the research community. Studies on web-mediated genres (e.g. Bondi et al., 2015; Bondi & Cacchiani, 2021) have developed special attention to the issue of making knowledge accessible also to non-experts (or experts in other fields). The process is often seen as one of reformulation of expert discourse (e.g. Gotti, 2014) and recontextualisation (Calsamiglia & van Dijk, 2004, p. 371). The expectation is also that knowledge will be used by the receiver to change practices or viewpoints or for intellectual growth (see also Bondi, 2020) in an active learning process aiming at an effective use of the transferred knowledge.

The elements defining the process are thus diverse, ranging from the content to be disseminated to the medium of communication, but above all, the context of the message and the intended users. The purposes may vary according to context: knowledge transfer in university–industry research partnership, knowledge translation from basic science to clinical trials in "bench-to-bedside" medical research, and so on. But also different reasons may guide people's interest in acquiring new knowledge, ranging from getting guidance in their practices, satisfying their curiosity, and pursuing intellectual growth to exploring or confirming viewpoints. And different purposes may guide scientists in their attempt to reach out to a wider audience, as shown quite clearly by Engberg (this volume) in his interesting attempt to map different levels and forms of engagement, ranging from forms of vulgarisation (entertainment) to more neutral dissemination of information about research, through to popularisation actually aiming at readers' empowerment. Beyond the binary distinction between research communication and popularisation, the digital transformation has highlighted a whole cline of possibilities that deserve further exploration. The chapters in the volume edited by Plo-Alastrué and Corona Marzol provide interesting insights into his cline of possibilities.

Università degli Studi di Modena e Reggio Emilia, Modena, Italy

Marina Bondi
marina.bondi@unimore.it

References

Baron, N. (2008). *Always on: Language in an online and mobile world.* Oxford University Press.

Berg, M., & Seeber, B. K. (2016). *The slow professor: Challenging the culture of speed in the academy*. University of Toronto Press.

Bondi, M. (2018a). Dialogicity in written language use: Variation across expert action games. In E. Weigand & I. Kovecses (Eds.), *From pragmatics to dialogue* (pp. 137–170). John Benjamins.

Bondi, M. (2018b). "*Try to prove me wrong*": Dialogicity and audience involvement in economics blogs. *Discourse, Context & Media, 24*, 33–42. https://doi.org/10.1016/j.dcm.2018.04.011.

Bondi, M. (2018c). Blogs as interwoven polylogues. *Language and Dialogue, 8*(1), 44–66.

Bondi, M. (2020). Academics online: Code glosses across research genres and public communication. In M. Gotti, S. Mack, & M. Sala (Eds.), *Scholarly pathways. Knowledge transfer and knowledge exchange in academia* (pp. 57–82). Peter Lang.

Bondi, M. (2022). Dialogicity in individual and institutional scientific blogs. *Publications, 10*(1), 9.

Bondi, M., & Cacchiani, S. (2021). Knowledge communication and knowledge dissemination in a digital world. *Journal of Pragmatics*, 186, 117–123.

Bondi, M., Cacchiani, S., & Mazzi, D. (Eds.). (2015). *Discourse in and through the media: Recontextualizing and reconceptualizing expert discourse.* Cambridge Scholars Publishing.

Calsamiglia, H., & Van Dijk, T. A. (2004). Popularization discourse and knowledge about the genome. *Discourse & Society, 15*(4), 369–389.

Cope, B., & Kalantzis, M. (2014). Changing knowledge ecologies and the transformation of the scholarly journal. In B. Cope, & A. Phillips (Eds.), *The future of the academic journal* (pp. 9–83). Chandos Publishing.

Corona, I. (2021). A window to the world: Visual design and research visibility of European research projects' homepages. In R. Lorés-Sanz, & G. Diani (Eds.), Disseminating knowledge: Digital discourse in academic and institutional contexts (Special Issue). *European Journal of English Studies 25*, 355–371.

Gotti, M. (2014). Reformulation and recontextualization in popularization discourse. *Ibérica, Revista de la Asociación Europea de Lenguas para Fines Específicos*, (27), 15–34.

Herring, S., Stein, D., & Virtanen, T. (Eds.). (2013). *Pragmatics of computer-mediated communication*. DeGruyter.

Holmberg, K., & Thelwall, M. (2014). Disciplinary differences in Twitter scholarly communication. *Scientometrics, 101*(2), 1027–1042.

Hyland, K. (2016). *Academic publishing: Issues and challenges in the construction of knowledge*. Oxford Applied Linguistics. Oxford University Press.

Hyland, K., & Jiang, F. K. (2019). *Academic discourse and global publishing: Disciplinary persuasion in changing times*. Routledge.

Johnson, R., Watkinson, A., & Mabe, M. (2018). The STM report. *An overview of scientific and scholarly publishing*. 5th edition October.

Jünger, J., & Fähnrich, B. (2020). Does really no one care? Analyzing the public engagement of communication scientists on Twitter. *New Media & Society, 22*(3), 387–408.

Kastberg, P. (2010). Knowledge communication. *Formative Ideas and Research Impetus. Programmatic Perspectives, 2*(1), 59–71.

Klar, S., Krupnikov, Y., Ryan, J. B., Searles, K., & Shmargad, Y. (2020). Using social media to promote academic research: Identifying the benefits of Twitter for sharing academic work. *PloS one, 15*(4), e0229446.

Lee, J. J., & Haupt, J. P. (2021). Scientific globalism during a global crisis: Research collaboration and open access publications on COVID-19. *Higher Education, 81*(5), 949–966.

Lillis, T. M., & Curry, M. J. (2010). *Academic writing in a global context*. Routledge.

Lockley, P., & Carrigan, M. (2011). Continual publishing across journals, blogs and social media maximises impact by increasing the size of the 'academic footprint'. *Impact of Social Sciences Blog*.

Lorés, R., & Diani, G. (Eds.). (2021) Disseminating knowledge. The effects of digitalised academic discourse on language, genre and identity. Special issue of the *European Journal of English Studies, 25*(3), 249–258. https://doi.org/10.1080/13825577.2021.1988262

Lorés, R. (2020). Science on the web: The exploration of European research websites of energy-related projects as digital genres for the promotion of values. *Discourse, Context & Media, 35*, 1–10.

Luzón, M. J., & Pérez-Llantada, C. (Eds.). (2019). *Science communication on the Internet: Old genres meet new genres* (Vol. 308). John Benjamins.

Marshall, P. D., Kim, B., & Christopher, M. (2017). Academic persona: The construction of online reputation in the modern academy. In D. Lupton, I. Mewburn., & P. Thomson, (Eds.), *The digital academic* (pp. 47–62). Routledge.

Mauranen, A. (2010). Features of English as a lingua franca in academia. *Helsinki English Studies, 6*(6), 28.

Mur-Dueñas, P. (2018). Disseminating and constructing academic knowledge in online scholarly journals: An analysis of virtual special issue introductions. *Discourse, Context & Media, 24*, 43–52.

Mur-Dueñas, P. (2021). Engagement markers in research project websites: Promoting interactivity and dialogicity. *Poznań Studies in Contemporary Linguistics 57*(4), 655–676.

Pascual, D., Mur-Dueñas, P., & Lorés-Sanz, R. (2020). Looking into international research groups' digital discursive practices: Criteria and methodological steps in the compilation of the *EUROPRO* digital corpus. *Research in Corpus Linguistics 8* (2), 87–102.

Pérez-Llantada, C. (2016). How is the digital medium shaping research genres? Some cross-disciplinary trends. *ESP Today 4*(1), 22–42.

Plo-Alastrué, R., & Pérez-Llantada, C. (Eds.). (2015). *English as a scientific and research language: Debates and discourses* (English in Europe, Vol. 2). Walter de Gruyter.

Richardson, E. T. (2020). Pandemicity, COVID-19 and the limits of public health 'science'. *BMJ Global Health*, *5*(4), e002571.

Vicente-Sáez, R., & Martínez-Fuentes, C. (2018). Open Science now: A systematic literature review for an integrated definition. *Journal of Business Research, 88*, 428–436.

Weitkamp, E., Milani, E., Ridgway, A., & Wilkinson, C. (2021). Exploring the digital media ecology: Insights from a study of healthy diets and climate change communication on digital and social media. *Journal of Science Communication, 20*, A02.

Yus, F. (2015), Interactions with readers through online specialized genres: Specificity or adaptability? In L. Gil-Salom, & C. Soler-Monreal (Eds.), *Dialogicity in written specialised genres* (pp. 189–208). Benjamins,

Zappavigna, M. (2012). *Discourse of Twitter and social media: How we use language to create affiliation on the web*. Bloomsbury.

Acknowledgements

This volume has been funded by the Spanish Ministerio de Industria, Economía y Competitividad under the projects "Toward Greater Visibility and Dissemination of Scientific Research" (FFI2017-84205-P) and "SciDis: Processes of Recontextualisation in the Transfer of Knowledge" (PID2021-122303NB-100).

We wish to express our warmest gratitude to our colleagues, the members of the InterGedi research group (Universidad de Zaragoza), for their continued support and helpful collaboration in the process of publication of this volume.

Contents

Notes on Contributors

Vicent Beltrán-Palanques is an assistant professor in the Department of English Studies at Universitat Jaume I and a member of the GRAPE research group (Group for Research on Academic and Professional English). His research interests include multimodal literacy, multimodal discourse analysis, ESP pedagogy and English-medium instruction. He has participated in national and international conferences. Part of his research has been published in journals like the *Journal of English Studies, System, Revista Alicantina de Estudios Ingleses, CALL-EJ (2021)* and *Journal of English for Academic Purposes*. He has authored book chapters published in international publishers such as *Routledge*, *Springer* and *Tirant lo Blanc*. He has participated in several research projects related to the field of multimodal discourse analysis and English-medium instruction, as well as in innovation projects.

Fatma Benelhadj is an assistant professor at the Faculty of Letters and Humanities of Sfax (FLSHS). She holds a PhD in English Linguistics from the same institution. She lectures mainly on English grammar, semantics, cognitive linguistics, translation theory, and academic writing. Her research interests include systemic functional linguistics, conceptual metaphor and translation. She is also a member of Laboratory on Approaches to Discourse (LAD), treasurer of SYFLAT—an association promoting research on systemic functional linguistics—and member of a project on gender equality (INSAF-FEM). She is one of the core members of the 'Pôle Étudiant Entrepreneur' (PEES), which aims to promote the entrepreneurial spirit of students. Moreover, she is the coordinator of RANKUS in the Faculty of Letters and Humanities of Sfax (FLSHS): a unit working on promoting the ranking of the University of Sfax. She has published, among others, in *Perspectives from Systemic Functional Linguistics: An Appliable Theory of Language* and in *Journal of Pragmatics*.

Vijay K. Bhatia is a retired Professor from City University of Hong Kong and is now Adjunct Professor at the Chinese University of Hong Kong and Visiting Professor at the Hellenic American University in Athens (Greece). Some of his research projects include *Analyzing Genre-bending in Corporate Disclosure Practices* and *International Arbitration Practice: A Discourse Analytical Study*, in which he led research teams from more than 20 countries. His research interests include (critical) genre theory, analysis of academic and professional discourses, particularly in legal, business, promotional and new media contexts; ESP and professional communication; simplification and 'easification' of legal and other public documents. Three of his monographs on genre analysis, *Analysing Genre: Language Use in Professional Settings* (1993), *Worlds of Written Discourse: A Genre-based View* (2004) and *Critical Genre Analysis: Interdiscursive Performance in Professional Practice* (2017) are widely used in genre theory and practice.

Ana Bocanegra-Valle is a senior lecturer at the University of Cádiz (Spain) where she teaches Maritime English at graduate and postgraduate levels. Her main research interests include Maritime English, ESP/EAP methodology, English for research publication purposes, and ESP/EAP discourse. She has been the editor-in-chief of the Languages for Specific Purposes (LSP) journal *Ibérica* for eight years and is at present Book Review editor for *ESP Today* and *Spanish Journal of Applied Linguistics*. She has published articles and book chapters in prestigious journals and publishing houses and has served as peer reviewer for a wide number of journals both in Spain and abroad. She is the head of the research group LaCAP (Languages Applied to Communication in Academic and Professional Settings). Her latest books are *Applied Linguistics and Knowledge Transfer: Employability, Internationalisation and Social*

Challenges (2020) and *Ethnographies of Academic Writing. Theory, Methods, and Interpretation* (co-edited, 2021).

Marina Bondi is Professor of English Linguistics at the University of Modena and Reggio Emilia. Founding Director of the *CLAVIER* centre *(Corpus and LAnguage Variation In English Research)*, she has published extensively in the field of genre analysis, EAP and corpus linguistics, with a focus on argumentative dialogue and language variation across genres, disciplines and cultures. Her recent interest centres on knowledge dissemination and the impact of digital media and digital communication on specialised academic and professional discourse. She coordinates the Modena Unit of a national project on *Communicating Transparency: New Trends in English-Language Corporate and Institutional Disclosure Practices in Intercultural Settings.*

Ruth Breeze is Full Professor of English at the University of Navarra, Spain, and PI of the Public Discourse Research Group in the Instituto Cultura y Sociedad. Her most recent books are *Pandemic and Crisis Discourse: Communicating COVID-19 and Public Health Strategy* (with Andreas Musolff, 2022), *Teaching English Medium Instruction Courses in Higher Education* (with Carmen Sancho Guinda, 2021) and I*magining the Peoples of Europe: Populist Discourses Across the Political Spectrum* (with Jan Zienkowski, 2019). Together with Maria Kuteeva, she is Editor-in-Chief of *Ibérica.*

Shirley Carter-Thomas is Professor of English linguistics and Communication at Institut Mines-Télécom Business School and a member of the French CNRS research team LATTICE. She is also a member of the French national ESP research association, GERAS, where she participates in the activities of the special interest group on academic literacy. Her research areas span functional and contrastive linguistics, genre analysis and the linguistic analysis of written and oral academic discourse. With Elizabeth Rowley-Jolivet, she has recently co-authored a series of articles focusing on the specificities of a number of web-mediated genres, including science podcasts, open lab notebooks, research group videos and three-minute thesis presentations on YouTube.

Isabel Corona is Senior Lecturer in English Studies at the Universidad de Zaragoza (Spain). Her research has focused on the analysis of academic and professional genres and discourses in traditional and new media contexts. She has participated in several national and international projects, on international commercial arbitration practices, on generic integrity in academic and professional communication, on English as a lingua franca across specialised discourses and on web-mediated visibility of scientific research. She has published in a number of international journals and edited volumes. She is engaged in the processes of multimodal recontextualisation in professional digital contexts.

Giuliana Diani is Associate Professor of English Language and Linguistics at the University of Modena and Reggio Emilia, Italy. She holds an MA in Language Studies from the University of Lancaster (UK) and a PhD in English Linguistics from the University of Pisa (Italy). She has worked and published on various aspects of discourse analysis and EAP, with special reference to language variation across academic genres, disciplines, and cultures through the analysis of small, specialised corpora. Her recent work has centred on the analysis of language use in knowledge dissemination addressing children and teenagers.

Olga Dontcheva-Navratilova is Associate Professor of English Linguistics at the Faculty of Education, Masaryk University, Czech Republic. Her research interests include English for academic and specific purposes and political discourse. She has published the books *Analysing Genre: The Colony Text of UNESCO Resolutions* (2009) and *Coherence in Political Speeches* (2011) and co-authored *Persuasion in Specialised Discourse* (2020). She is co-editor of the journal *Discourse and Interaction.*

Jan Engberg is Professor of Knowledge Communication at the Department of German Business Communication, School of Communication and Culture, Aarhus University, Denmark. His main areas of research interest are the study of texts and genres in the academic field, cognitive aspects of domain-specific discourse and the relations between specialised knowledge and text formulation as well as basic aspects of communication in domain-specific settings. His research has been focused upon communication and translation in the field of law as well as other fields of academic communication like climate change communication. He has widely published his investigations in high-quality journals such as *Journal of Pragmatics, Parallèles* or *International Journal of Legal Discourse.* He is furthermore editor of the international journals *Fachsprache* and *Hermes* and member of advisory boards of a considerable number of international scientific journals like *Ibérica* and *Comparative Law and Language.*

Maria Freddi is Associate Professor of English Language and Linguistics at the University of Pavia, Italy, where she has been teaching corpus linguistics and ESP/EAP courses with a focus on English for science and technology and academic and popular writing. From 2019 to 2022, she was a member of the Erasmus+ Becoming a Digital Global Engineer, BADGE, research project, aimed at developing open language resources for students of engineering around Europe. She is working on a book for John Benjamins, a monograph on writing in science and technology.

Juan C. Palmer-Silveira, PhD in English Philology, is a member of the research group GRAPE (Group for Research on Academic and Professional English) and is working as an associate professor at Universitat Jaume I (Castelló, Spain), where he is also coordinating the University Master in English Language for International Trade (ELIT), currently in its 18th edition. His main research fields are English writing techniques, as well as English for professional and academic purposes, having specialised in English for Business Communication. In recent times he has been working on multimodal discourse analysis (MDA). He has published in relevant international journals (*Comunicar, RELC, Ibérica, Business and Professional Communication Quarterly, The Reading Matrix, British Journal of Educational Technology*) and in well-known publishing houses (Peter Lang, Rodopi, Farleigh Dickinson, Routledge, John Benjamins, Narcea).

Daniel Pascual holds a PhD in English Studies from the Universidad de Zaragoza (Spain). His doctoral thesis focused on the pragmatic strategies employed in digital scholarly practices by international scientific research groups for knowledge dissemination and e-visibility, such as websites and social media. As part of the InterGedi research group, his research interests comprise digital pragmatics, digital discourse analysis, multimodality, metadiscourse and ethnographically oriented analyses. He has published his work in a number of national and international, high-quality journals.

Ramón Plo-Alastrué is Professor of English Linguistics at the Universidad de Zaragoza (Spain). His interests include the analysis of traditional written academic genres (i.e. patterns of lexicogrammatical, rhetorical and discursive features); the identification of cultural norms and conventions adhered to in institutional research in the academic 'semiperiphery'; and, as a member of the InterGedi research group, digital genre analysis.

Elizabeth Rowley-Jolivet is a research member of the Laboratoire Ligérien de Linguistique, University of Orleans, France, and of the French national ESP research association, GERAS. Her research covers the multimodal and linguistic features of spoken and written scientific discourse, genre analysis, academic literacies and the epistemology of science. Her interest in genre evolution and in Open Science led her to turn to digital genres in the early 2010, when she co-convened two ESSE seminars on *From Downloading to Uploading: New Spaces and New Voices in Web Authoring* (ESSE 10, 2010) and *From Print to Web 2.0 : What Future for professional discourses?* (ESSE 11, 2012) and co-edited the volume *Evolving Genres in Web-mediated Communication* (Peter Lang, 2012). She has since then published studies on open science laboratory protocols, open lab notebooks, science podcasts and videos and three-minute thesis presentations.

Miguel F. Ruiz-Garrido, PhD member of the research group GRAPE (Group for Research on Academic and Professional English), is an associate professor at Universitat Jaume I (Castelló, Spain). His main research fields are English for professional and academic purposes, including English for Business Communication or Multimodal Discourse Analysis, among others. He has published in international journals (*Business and Professional Communication Quarterly, Ibérica, ESP Today, British Journal of Educational Technology, System* or *Revista de Lenguas para Fines Específicos*) as well as in well-known publishing houses. His most recent publications are "Examining Students' Reflections on a Collaborative Online International Learning Project in an ICLHE Context" (Mestre-

Segarra and Ruiz-Garrido, *System*, 2022) and "Moodle E-learning Platform as a Complementary Tool in ICLHE Contexts" (Ruiz-Garrido and Fortanet-Gómez, *RLFE*, 2022).

Noelia Ruiz-Madrid is a senior lecturer at Universitat Jaume I, Spain, where she lectures English writing, English language learning and teaching, and digital genres at graduate and postgraduate levels. Her research interests are MDA (Multimodal Discourse Analysis), academic and professional genres, and digital genres. Her publications have appeared in *Ibérica*, *Discourse Studies*, *International Journal of English Studies*, *Language and Communication* and *Journal of English for Academic Purposes*.

Carmen Sancho-Guinda is a senior lecturer in the Department of Applied Linguistics at the Universidad Politécnica de Madrid, Spain, where she teaches English for Academic and Professional Communication and in-service seminars for engineering teachers undertaking English-medium instruction. Her research interests comprise the interdisciplinary study of academic and professional discourses and genres. Among her publications in these areas are *Stance and Voice in Written Academic Genres* (2012, coedited with Ken Hyland for Palgrave), *Narratives in Academic and Professional Genres* (2013, coedited with Maurizio Gotti and winner of the 2015 Enrique Alcaraz Research Award), *Interpersonality in Legal Genres* (2014, coedited with Ruth Breeze and Maurizio Gotti), *Essential Competencies for English-Medium University Teaching* (2017, coedited with Ruth Breeze for Springer), *Engagement in Professional Genres* (edited in 2019), and *Teaching English-Medium Instruction Courses in Higher Education* (2021, co-written with Ruth Breeze). She was editor-in-chief of *Ibérica* from 2018 to 2021 and is an editorial member of the *Journal of English for Academic Purposes* (Elsevier), *Frontiers in Communication, Revista de Lingüística y Lenguas Aplicadas,* and the University of Bergamo's CERLIS series.

Julia Valeiras-Jurado is an assistant professor at Universitat Jaume I, Spain, where she lectures English for Specific Purposes and English writing. Her research interests are multimodality, academic and business discourse, academic and business genres, oral genres and persuasive language. Her publications have appeared in *Ibérica*, *Text & Talk*, *International Journal of English Studies*, *Language and Communication* and *Journal of English for Academic Purposes.*

List of Figures

List of Tables

Part I

An Introduction to Scientific Research Communication Through Digital Media

1

Digital Scholarly Practices in Scientific Communication: Paths and Goals in Research Dissemination

Daniel Pascual, Ramón Plo-Alastrué, and Isabel Corona

1 Introduction

The mutual entente between the use of the Internet and the development of society is undeniable. As users, we take advantage of the digital medium to enact actions and pursue goals in our daily communicative encounters. The traditional dichotomies of writer versus reader, when looking into texts, and producer versus consumer, when dealing with material goods, have merged at an unequalled scale in the discourses constructed online. The infrastructure of the Web enables users to blur the boundaries around the role they play in digital communication. The duties of creation, modification, distribution and consumption of information are accordingly shared by the whole Internet community, leading to its members being regarded as 'produsers' (Bruns, 2007).

D. Pascual (✉) • R. Plo-Alastrué • I. Corona
Universidad de Zaragoza, Zaragoza, Spain
e-mail: dpascual@unizar.es; plo@unizar.es; iscorona@unizar.es

R. Plo-Alastrué, I. Corona (eds.), *Digital Scientific Communication*,
https://doi.org/10.1007/978-3-031-38207-9_1

Beyond personal and private spheres, it is remarkable how digital communication has expanded in professional environments, which have appropriated its affordances for commercial, corporate, advertising, pedagogical, entertainment and research purposes. Of special relevance is the paradigm change the digital medium has triggered for scholars and scientists. Parallel to the establishment of Web 2.0, characterised by the participatory nature of the platforms and media available to users, the notions of 'Scholarship 2.0' and 'Science 2.0' unveil new approaches in the construction and communication of knowledge, based on the cornerstones of collaboration, network technologies, accessibility and information sharing. The rapidly increasing interests of institutions, universities, research centres and associations in occupying a prominent position at the forefront of knowledge creation and research dissemination may be pushing scholarly and scientific practices towards the commodification of information and the marketisation of science. Yet, the negative implications that could derive from this view are countered with the perks of a post-modern globalised society, which also aspires to the democratisation of knowledge and the participation of citizenship.

Consequently, the ways of *doing* and *talking about* scientific research continue to flourish, becoming more varied, complex and ambitious. Research has recently attempted to depict the new genres and practices employed under the umbrella term of science communication (e.g. Bartling & Friesike, 2014; Bucchi & Trench, 2016; Davies & Horst, 2016; Luzón & Pérez-Llantada, 2019; Lorés & Diani, 2021; Mur-Dueñas & Lorés, 2022, to name but a few). More specifically, the premise of this volume is to gain understanding of current practices in 'scientific communication', which concerns communication among equals, encompassing a variety of agents in the scientific process that includes citizens, as opposed to 'science communication', which rather addresses the unidirectional communication from scientists to non-scientists (Zerbe et al., 2022).

2 Scientific Communication: Models, Approaches and Practices

A critical approach to 'science communication' is needed, since its conceptualisation very much depends on the period in time and the contexts in which it has been used. The changes in 'science communication' throughout time have resulted in three principal conceptual models to describe the shifting directions that it has experienced in its pursuits, target publics, channels of communication and favoured ways of expression. As it may be hardly observable how agents in science communication practices have internalised these models, they should therefore be regarded probably as coexistent rather than replacing one another. In the 'Deficit Model', associated with the 1960s and giving way to the Public Understanding of Science movement in the 1980s and 1990s, laypeople are assumed to have a deficit of scientific understanding that should be repaired. An increase in the knowledge and literacy of the public is needed, which may drive to more support and legitimation of science on their part. Thus, a top-down, one-way view on scientific knowledge is contended, foregrounding "the public's inability to understand the achievements of science—according to a model of linear, pedagogical and paternalistic communication" (Bucchi & Trench, 2016, p. 155). The problem is attributed to the public rather than to science itself or to scientists in their efforts to communicate their findings. Hence, science communication is characterised by its simplicity and easy implementation, and places the public away from influencing scientific discourse. At present, it would be represented by mediated divulgation where the public is taken to be persuadable and with no judgement in scientific communication processes.

In the 2000s, the 'Dialogic Model' gained momentum to facilitate the Public Engagement with Science, bringing about a two-way dialogue between science and citizens. As a result, a greater value was placed on the interaction between academics and stakeholders. This is the point at which the dichotomy between 'science' and 'scientific' communication finds its roots, prioritising a process shared among equals. To attain the involvement of non-scientific agents in the processes of knowledge production and scientific investigation, a transition is imperative from awareness into involvement, from communication into dialogue and overall from Science *and* Society to Science *in* Society (Bucchi & Trench, 2016). Although public engagement is fostered, critical and sceptical views may remain on the grounds that the dialogue in this model is not genuinely symmetrical (Engberg, this volume). The original producers of science and its communication take over, while citizens still have no significant role in the outcomes produced. Scientists, thus, aim at communication in authentic contexts and at attentive listening to citizenship, but in the end, they direct the exchange of views between laypeople and experts towards the latter, due to perceived knowledge deficits of the former.

As the most recent paradigm shift, the 'Participation Model', revolves around the notions of Open Science, Citizen Science, Strategic Science Communication and Upstream Public Engagement. Participation is seen as "a stronger form of engagement by the public both with scientific ideas and with the governance of science" (Bucchi & Trench, 2016, p. 157). The focus is then placed on the legitimation of scientific protagonists and the increased autonomy of institutions and individuals to promote science and its dissemination, bringing meaningful implications for scholars and citizens. The instrumentality of scientific communication is a prominent feature in this model, as the process of participating in knowledge distribution brings along reputation, credibility, impact, image-building, self-branding and positive identity construction. Subsequently, active agents are given now the possibility of participating in the scientific processes taken by professionals themselves and negotiating scientific knowledge with them. Research projects and activities count on the participation of non-scientific agents, who help generate new scientific knowledge and findings. This is accompanied by a movement towards Open Access (OA) to outlets in which results can be freely and widely consulted. As Kessler et al. (2022, p. 17) contend, "[T]he increasing number of scientists interacting with the media and conducting other public engagement activities may in itself be changing the culture of science." The conception of Science *in* Society has thus further evolved to tackle the development of Science *with* Society.

In light of the axioms in these models, scientific information and results are being exponentially communicated to ensure a more democratic knowledge and participatory culture. The previously dominant, traditional view of translating, simplifying and distorting scientific discourse, as though scientists were constitutive authorities and the public was absolutely ignorant of scientific matters, seems to be rejected. As such, the unidirectional transmission of knowledge from experts to citizens is surmounted and a wider communication is embraced, which trespasses institutional and personal networks and which credits other means and agents of disseminating knowledge among scientists and researchers. Puschmann (2015, p. 26) affirms that "rather than merely making the fruits of scholarly research available to the public, citizens are increasingly regarded as active stakeholders in the scientific process", to which Lorés (2020, p. 8) adds that "scientific discourse is taking shape in a context of *civil science*, where policies are increasingly undertaken to make science accessible to the public, whether expert or not". The enhancement of the openness to science does not only entail widening its access to citizenship, but also boosting their interest in scientific endeavours. In this regard, the ubiquity and immediacy intrinsic to digital communication are essential. Such a change of mind reflected in still-to-be-explored opportunities within the online medium to build communicative bridges among audiences of diverse levels of specialisation nurtures the contributions of this edited volume.

Concerning the contexts of situation, 'science communication' may imply very different conceptions for different societal agents. Science researchers may conjure up exclusively academic endeavours in which they write scientific papers and speak at in-group events; science journalists may elicit images in which they mediate the information retrieved from researchers to offer it adapted to the public; science communicators may think of tailoring stimulating activities and raising awareness and interest about science in informal educational settings; and science teachers may focus on didactic perspectives that develop trainees' skills and literacies through written and oral works and foster formative assessment among peers using dialogue and teamwork to assimilate knowledge (see a practical model in Beltrán-Palanques, this volume). These approaches illustrate the variety of mindsets that influence people to communicate science.

An amalgam of competing, yet interrelated, terms (i.e. 'dissemination', 'popularisation', 'vulgarisation') has boomed and should be acknowledged to reflect these realities in the current panorama of research transfer, especially as amplified by the potential of digital practices. They all serve to depict the manifold ways in which research is transmitted and shared, pursuing various goals and manifesting dissimilar relationships between researchers and the audience. Some of them are at present more in vogue than others, responding to the democratising values and the engaging efforts associated with research endeavours. Drawing on the framework of Knowledge Communication (e.g. Kastberg, 2010; Ditlevsen, 2011), these concepts enable us to analyse and comprehend "the intentional and decision-based communication of specialised knowledge in professional settings [...] with a focus upon the interplay between knowledge and expertise of individuals, on the one hand, and knowledge as a social phenomenon" (Engberg, 2016, p. 37).

'Dissemination' may be the most neutral and commonly favoured notion in scientific communication with a focus on knowledge content and a clear reference to the circulation and delivery of academic information and research findings. By contrast, 'popularisation', defined as "science writing for the general public" (Calsamiglia, 2003, p. 139), puts the emphasis on building bridges, often through emotional bonds, among research producers and consumers. In this sense, both terms may involve expert-to-(non)expert communication, but popularisation leans much more upon the didactics of knowledge, typically in institutionalised contexts. In a broad sense, popularising practices involve, then, a recontextualisation of information in such a way that it is more comprehensible and pertinent to various audiences. In a more restricted view, "popularization includes only texts about science that are not addressed to other specialist scientists" (Myers, 2003, p. 265), as if there were two separate discourses with clear boundaries. Two more concepts derive from the notion of 'popularisation', as they alter the purposes and relationships between researchers and their audience. 'Vulgarisation' refers to the social appropriation of the public understanding of science and technology and entails discursive processes to diminish the level of specialised information to something popular, familiar and ordinary to the reader. Moreover, 'infotainment' or 'scifotainment' has also sprung up as a popularising approach, which concentrates on entertainment material based on domains of popular culture (Zhang, 2019). Thus, it seeks to engage the readership and nurture committed relationships with them and rather leaves scientific information as a spin-off (Engberg, this volume). It would not be fair to leave out other labels that also portray the reality of scientific information approaches and processes. Some of them comprise 'disinformation' and 'misinformation' (Bhatia, this volume), 'trivialisation', 'parascience', 'pseudoscience' and 'speculation' (as in predatory practices;

see Bocanegra-Valle, this volume), and help complete the fuzzy picture of the revolution caused by the digital medium in knowledge distribution and research dissemination.

The emphasis in response to all these terms characterising the communication of science differs among individuals, professional disciplines, geographical locations and sociocultural backgrounds. It is essential for researchers and trainers in scientific communication to establish shared knowledge with their audiences, whether stakeholders or non-specialised readers. There are some central terms that play an undeniable role in the directions science communication is taking and the goals researchers seek to accomplish, which we discuss in the following section.

3 Key Concepts in Research Dissemination

The community of professionals concerned with the pursuit of research, education and scholarship (to whom we generally refer collectively as 'the academia') has seen how emerging digital realities present new challenges in their discursive and professional practices. Traditional written genres, such as the research article, to name but the conventional academic genre *par excellence*, need to be reshaped in order to adapt to new platforms. Genres native to the digital medium, such as blogs or video abstracts, pose new production demands for authors while opening the research space to innovative dissemination perspectives, as reception expands to multiple audiences, who move from being just readers to active interactants, that is, participants in knowledge dissemination. The challenges and consequences of research dissemination in the digital medium bring along the attempt to achieve a higher impact of one's work. Two aspects seem to contribute to that end, namely, constructing a solid digital identity and ensuring an effective visibility of research. The opportunities provided by the affordances of the digital medium and affected by the elusive notion of audience should be borne in mind when handling research dissemination online. In what follows, we discuss four main concepts at the core of the trends in scientific communication, in general, and research dissemination, in particular.

3.1 Digital Communication

Digital communication is at the forefront of all the spheres of our lives. As such, the analysis of digital discourse has become a prominent research area to investigate the interplay of new media and social uses of language, including metalanguage and language ideology (Thurlow & Mroczek, 2011), the evolution of traditional genres into new digital hybrids (Jones et al., 2015), the construction of users' face and identity through multi-semiotic resources (Bou-Franch & Garcés-Conejos Blitvich, 2019), as well as the rise of social media channels in situated communicative events and practices (Johansson et al., 2021). These concerns are naturally extensible to the dissemination of science and research, which go "hand in hand with technological development and, in general, with the need to apply scientific advancements to the improvement of human wellbeing" (Mur-Dueñas & Lorés, 2022, p. 1).

The affordances of the digital medium have a self-evident impact in the way we consume and generate new content in general, and specialised discourse, in particular. Texts produced in digitally mediated communication are characterised by their non-linearity, giving way to manifold "navigating paths" (Askehave & Nielsen, 2005) and making use of "sites for action" (Adami, 2015). In the same vein, they are inherently interactive and multimodal, comprising a combination of modes (e.g. verbal, visual, auditory) rhetorically organised to enquire "how the resources and processes of meaning-making shape and are shaped by people, institutions and societies" (Jewitt et al., 2021, p. 4).

This digital landscape has been triggering enormous consequences for scholarly work and professional communication, amplifying researchers' intended actions and enhancing the dissemination of knowledge. It is worth, then, investigating how the epistemic authority of science is at a turning point thanks to digital, multimodal communication, privileging the democratic access to information and the co-construction of knowledge. The affordances at researchers' disposal are combined in scientific communication not only to accomplish their intended actions through more and more complex meaning-making processes but also to comply with their growingly sophisticated purposes of informativity, visibility, reputation, identity, self-branding and interaction. Thus, some immediate consequences for the research world lie in the preponderance of social media, the pursuit of speedy publication, wider dissemination of preprints and ethical questionable research practices (Bhatia, this volume).

We, as scholars, utilise digital affordances to particular situations, relationships and identities, negotiating what we can think, do and mean as much as how we can relate to others (Jones & Hafner, 2012). Overall, the analysis of trends in digital scientific communication allows us to pinpoint current practices, emergent objects of study and socioeconomic and institutional realities which altogether affect how research dissemination is encouraged and researchers' visibility promoted.

3.2 Identity

Anyone's identity is socially framed—full of individual complexities and in permanent evolution—and entails a selective self-presentation out of performance (Goffman, 1959). The notion of 'identity' should then be considered as a polyhedron with manifold sides that we construct, decorate and display based on personal, social and communicative experiences. As an intangible reflection of our self and an organising feature that permeated all spheres of our everyday life, we can assume that our identity is publicly manifest to others through (digital) discourse in such a way that we display a "networked self" (Papacharissi, 2011).

In professional contexts, the comprehension of scholars' identity has also increasingly attracted attention (cf. Borgman, 2007; Bhatia & Evangelisti, 2012; Flowerdew & Wang, 2015). In the notion of digital identity, any user understands a degree of socialisation and the exposure of their face. In research environments, at the juncture of these two issues, lie particular facets, such as reputation, credibility, accountability, leadership and competitiveness. Researchers seek to curate their digital profiles through performative actions that combine instrumental and expressive goals. Their digital face, which grows apart from their offline counterpart, enables them to build a coexistent in-group persona within given professional communities and to raise their self-projection towards the wide public. Consequently, researchers target a heterogeneous set of users and a plethora of blurry contexts, for which they play out identities that cannot be construed in simplistic and mutually exclusive terms, that were possible earlier, such as online/offline, textual/embodied and fictional/authentic. Instead, all these characteristics fuse in a mutually constitutive way.

When users, researchers in this case, produce digital discourse, they share a representation of their 'self' with the audience through a dialectical relationship and a subsequent rhetorical performance (see, by way of instance, Rowley-Jolivet & Carter-Thomas, this volume). Researchers may take in the role of curators who select what to publish and how to do it to craft a coherent narrative of the self. They textualise themselves through carefully managed practices which are not to be taken as a transparent representation of reality, but as shaped by the way online interaction is enabled and valued for research dissemination purposes.

The ensuing complex scenario for scholars' everyday work has triggered significant changes in the construction of their personal and professional identities. The notion of identity is no longer approached as something fixed or static, but "as a fluid, plural accomplishment that is constantly under negotiation rather than a single, stable and essentialist entity" (Page, 2016, p. 403). Researchers of all disciplinary backgrounds need nowadays to *adopt* an ongoing set of practices and behaviours that may be unfamiliar to them and *adapt* previous assumptions and ways of doing to the new socioeconomic, political and digital reality that they are facing. Therefore, the notion of multifaceted identity seems to illustrate the many domains researchers are forced to excel in order to get promoted in their corresponding institutions, including teaching duties, carrying out investigations, engaging in projects, publishing high-quality papers and networking with other universities and scholars. Such a multifaceted identity gains an extra dimension in digital environments, where users feel compelled to be recurrently active in the digital sphere, and demonstrate online the many things they are constantly involved in. Digitally mediated communication seems to conform two sides of the same coin, where one side entails the inescapable endorsement of digital practices, skills and literacies to abide by the rules of the present scholarly system, whereas the other side brings along the potential establishment of an identity characterised by reputation and credibility, resulting from the fruitful combination of these facets and the appropriate exploitation of digital resources. In fact, the development of a personalised brand of the self for professional enterprises steers researchers to maintain contact with complementary audiences other than the strictly institutional and scholarly ones.

3.3 Visibility

Confronted with the traditional publishing formats widely recognised within academia (e.g. abstracts, research articles, edited volumes, conference presentations), researchers feel the urge to embrace new practices that may supplement them. The rationale behind this urge lies in the interest in bridging the potential gaps between their work and the general public at which their work may be directed. Consequently, the objectives of the publication and consumption of scientific communication have increased and publication in the traditional format has branched out into digital environments. The discourses produced by researchers and the interactions with users to exchange knowledge may seek to focus on the transfer of data, the release of fresh output, the facilitation of complex knowledge, intellectual growth, guidance and curiosity. Eventually, the rise of the digital medium has brought about researchers' aspiration for a higher visibility that may allow them to reach a wide audience, have a bigger impact and develop a digital identity that makes their investigations recognisable and prominent.

The potential of the Internet infrastructure to afford the distribution of content makes it possible to self-promote researchers' work on a large scale. According to Page (2012, p. 182), "visibility and attention have emerged as core properties necessary for accruing status and perceived influence." In all, using digital genres and social media to display one's research and promote the visibility of the researcher is regarded as an additional facet of the researcher's work duties and digital identity, parallel to the still prevalent conventional publication system. As Lorés-Sanz and Herrando-Rodrigo (2020, p. 84) claim, such digital visibility, or else "e-visibility", is "shaped by information generated by the researchers themselves and by others', and also by the context of the researchers' network".

It is then deemed necessary to regard researchers' visibility as polyphonic. The expanding wealth of digital practices at researchers' disposal accompanied by the influence of digital factors such as interactivity and altmetrics are revolutionising the way visibility is accomplished within the scholarly system. As such, fragmentation and pluralisation play a crucial role, in that visibility is accrued by developing manifold digital practices that complement each other in diverse ways in raising researchers' visibility (see, as an example, Dontcheva-Navratilova, this volume). As a result, this negotiation of visibility is permeated by processes of interdiscursivity, where promotional, corporate and academic discourses blend together for a positive representation of scholars and scientists. In the end, the visibility of research(ers) through digital practices entails a multivoiced perspective that combines explicit and subtle demonstrations of one's work and skills. To achieve e-visibility, researchers need to cater both for the informational gaps which separate them from digital users and for the interactional bridges that may bring them closer to such digital users. The problem may lie in the neat identification of the audience reached by researchers on their digital communication and dissemination practices.

3.4 Audience

Part of the reason for researchers to be conscious of the need to build a multifaceted identity and attain visibility of their work is linked to the slippery notion of 'audience' in digitally mediated communication. The fact that scholars address potentially global, diversified audiences in research dissemination makes the management of their multifaceted identities quite complex. This entails the endorsement of a growing span of digital practices, materialised in the use of novel genres and social media. An ample list of adjectives is usually considered to characterise digital audiences for research dissemination practices, which are undefined, global, general, varied, heterogeneous, diversified, targeted, imagined, real. Therefore, scholarly practices are imbued in polycontextuality, and the need to accommodate wider audiences is linked to a malleable continuum with eroding boundaries that may comprise lay citizens, interested readers, amateur-experts, expert audiences and professionals of the same and other disciplinary communities (Pascual & Mur-Dueñas, 2022).

Raising awareness about the potential addressees of the texts and genres in which research is encapsulated and disseminated is a relevant step that needs corresponding appropriate discursive and linguistic choices. Such metalinguistic awareness is necessary to handle the range of platforms and practices available online for researchers to establish outspreading interactional networks more than ever imaginable. Hence, heterogeneity is inevitably present in the likely readership of what researchers publish online, involving specialised users as much as the lay public that happens to visit and consume the published content. As a result, "science and the public are both much more diversified and fuzzy than their neat rhetorical separation would make us believe, and their relationship has a much less hierarchical character than in the bygone era of big science" (Puschmann, 2015, p. 33).

These ideas underline the importance and difficulty in understanding the audience in digitally mediated environments, in general, and within scholarly ones, in particular. The aftermath is that boundaries dividing professional and private networks are extensively fading. The digital identity portrayed by researchers, as affected by the likely uncertainty of the audiences targeted and the multifacetedness of their work, may be perceived as blurred, since it meshes private, personal and social traits to some extent. Such an unpredictability of the identities and characteristics of the audience to be reached has been explained through the notion of 'context collapse', especially due to the participatory and networking conditions of Web 2.0. In digital communication, context collapse is foregrounded because multiple audiences are flattened into one (Marwick & boyd, 2011), placing users from social, professional and public spheres all in the same boat. Following the dichotomy contended by Ito et al. (2010), as opposed to "friendship-driven practices", which imply dominant and mainstream practices in our everyday communicative events with our social groups, the focus of this volume is on "interest-driven practices", which enable socialisation with networks of peers and other participants based on specialised knowledge and scholarly interests.

In light of the topics discussed so far, it is clear that scientific communication can, and should, be analysed from many different angles. One of them concerns the large number of objects of study that can be investigated in digital research dissemination. The reader of this volume will find out analyses of online newspaper comments (Chap. 2, Breeze), technology disclosures (Chap. 3, Sancho Guinda), online magazine popular articles (Chap. 4, Benelhadj), law and science blogs (Chap. 5; Diani & Freddi), predatory e-mail messages (Chap. 6, Bocanegra-Valle), video abstracts (Chap. 8, Dontcheva-Navratilova), popular science online videos (Chap. 9, Ruiz-Madrid & Valeiras-Jurado), three-minute thesis presentations of different nature, including PhD thesis presentations (Chap. 10; Beltrán-Palanques; Chap. 12, Rowley-Jolivet & Shirley-Thomas) and FameLab scientific talks (Chap. 11, Palmer-Silveira & Ruiz-Garrido), and research group videos (Chap. 12, Rowley-Jolivet & Carter-Thomas).

Not only are the objects of study in research dissemination diverse and growing, but the methodological approaches to untangle information processes are also numerous and work many times in complementary ways. Ultimately, it is the sum of diverse analytical standpoints that may provide us with a holistic picture of the digital practices that are gaining momentum in the current panorama of scientific communication. The wide array of perspectives that applied linguists and discourse analysts endorse is carefully reflected in this volume. The reader will find analytical proposals which include reflective contributions on current scientific phenomena (Chap. 7, Engberg) and challenges and directions in publication and dissemination practices (Chap. 13; Bhatia), corpus-based analyses (Chap. 2, Breeze; Chap. 3, Sancho Guinda; Chap. 4, Benelhadj), mixed-method approaches (Chap. 5, Diani & Freddi; Chap. 8, Dontcheva-Navratilova; Chap. 9, Beltrán-Palanques; Chap. 12, Rowley-Joliet & Shirley-Thomas) and qualitatively oriented studies (Chap. 6 Bocanegra-Valle; Chap. 8, Ruiz-Madrid & Valeiras-Jurado; Chap. 10, Palmer-Silveira & Ruiz-Garrido). These perspectives are combined with further analytical frameworks which allow the contributors in this volume to offer well-rounded proposals on the research of digital practices, such as Genre Analysis, Conceptual Metaphor Analysis, Metadiscourse Analysis, Knowledge Communication Analysis, Multimodal Discourse Analysis and Thematic Analysis.

Furthermore, to conduct these diverse analyses, contributors have opted both for a close reading of their data in order to map fine-grained trends in research dissemination and for a number of software programmes which allow for more holistic analysis and help ensure methodological consistency. These two options attest the never-ending necessity of manually investigating discourse in situated contexts of production as well as the increasingly pertinent role of ICTs in our own research endeavours. Together with manual analyses (Chap. 3, Sancho Guinda; Chap. 4, Benelhadj), a spectrum of software tools is covered, comprising SketchEngine (Chap. 2, Breeze), AntConc (Chap. 5, Diani & Freddi), NVivo (Chap. 6, Bocanegra-Valle), MMAV (Chap. 9, Ruiz-Madrid & Valeiras-Jurado) and ELAN (Chap. 10, Beltrán-Palanques).

4 An Overview of the Chapters of the Volume

Drawing on a selection of papers presented at the InterGedi International Conference held in Zaragoza (Spain), December 2021, the common goal of the chapters included in this volume is to analyse some of the emerging professional practices in digital scientific communication, which make use of innovative multimodal genres and media, and explore their effects through a series of 'guiding' concepts, such as validated knowledge, visibility, dissemination and popularisation. Individual chapters in the resulting volume, *Digital Scientific Communication: Identity and Visibility in Research Dissemination,* have been organised into five parts:

Part I, **"An Introduction to Scientific Research Communication Through Digital Media"**, opens with "Prologue: State of the Art of Research Dissemination", where **Marina Bondi** addresses the changes triggered by a whole range of emerging digital practices on the traditional system of academic publication. In particular, she focuses on the way this impact brings to the fore—and often problematises—notions such as the author's identity and visibility as well as the changing roles adopted in the dissemination of knowledge, which are central to the studies presented in this book.

The present introductory chapter **1**, "Digital Scholarly practices in scientific communication: Paths and goals in research dissemination", co-authored by **Daniel Pascual** and the editors of this volume, **Ramón Plo-Alastrué** and **Isabel Corona**, draws attention to the coexisting models, approaches and practices that are gaining relevance in research dissemination. The authors introduce and revisit key concepts that permeate the analyses undertaken and presented in this volume. The plurality of approaches has been grouped according to similarities in their objects of study, methodological perspectives or analytical proposals, so as to articulate the structure of the volume into the following sections.

Part II "Scientific Discourse and Professional Practices" tackles the consequences of the digital turn in the production and reception of scientific content. In **Chap. 2**, "'Not One of Our Experts.' Knowledge Claims and Group Affiliations in Online Discussions of the COVID-19 Vaccine", **Ruth Breeze** explores the nature of public debate on scientific issues; she takes as object of study the 'reader comments' genre, one of the first genres to develop the interactional capacity of the digital media landscape. Taking a large sample of comments in response to articles about vaccines from the UK right-wing popular newspaper *Mail Online* published during the early phase of COVID-19 in 2020, Breeze adopts a discourse analytical perspective to identify the discursive strategies used by commenters to ground their authority, make claims about science and construct their identities. The analysis shows that commenters fell persistently into two polarised groups, pro- and anti-vaxxers, and built their arguments in two different ways: pro-vaccine participants invoked experts or adopted an expert-like style, while those against vaccines challenged expertise and prioritised personal experience and "common sense". Group-building strategies were also different. Intragroup dynamics of anti-vaxxers showed strong cohesion, emotional support and even humour and irony, while pro-vaxxers were less personal and less colourful. According to the author, this may be due to the affordances provided by the medium and the genre itself, as facilitators of confrontation. In the wider context, Breeze sees the clear erosion of expert knowledge as linked to ideological factors, and as a negative consequence of dissemination.

Carmen Sancho-Guinda's Chap. 3, "Utmost Hybridity: Promotional Trends in Technology Disclosures", focuses on this still under-researched digital genre of technology disclosure for technology dissemination. Stemming from the current context of growing academic and professional interactions brought about by digital affordances, technology disclosures are identified and approached in this chapter within a Genre Analysis framework at a co(n)textual, semiotic and verbal level. The resulting analysis provides a detailed picture of this evolving genre revealing its hybrid nature as one of its main distinctive features. In Sancho-Guinda's view, hybridity not only lies at the heart of those practices and promotional strategies which help configure discursive identity in technology disclosures but extends to the construction of an institutional identity, where the important role played by individual authors, "who make decisions as to format, content, conventions and connections", is also emphasised.

In **Chap. 4**, "Dissemination of Knowledge During the COVID-19 Pandemic: A Conceptual Metaphor Analysis of Research and Popular Articles", the use of metaphor to simplify new ideas and challenging concepts for the non-specialist reader is the starting point of analysis. **Fatma Benelhadj** compares a corpus of medical research articles on the COVID-19 pandemic and their corresponding popularisations, focusing on the specific functions of the most frequent metaphors in both genres. Drawing on Halliday and Matthiessen's (2014) ideational, interpersonal and textual metafunctions and on Semino's (2008) metaphorical Source Domains in medical science, Benelhadj reveals how this process of recontextualisation determines certain choices and popularising features, such as the common use of metaphors in the text. She delves into some of the functions performed by metaphors in popularisations (i.e. build the cohesion of the text, increase credibility and offer a space for researchers to express their identities) as well as in their genre-driven nature.

"Authorial Stance and Identity Building in Weblogs by Law Scholars and Scientists" is the title of **Chap. 5** by **Giuliana Dani** and **Maria Freddi.** The authors focus on the notion of stance and its contribution to the construction of identity in academic blogs. In their analysis, they combine corpus methodology and discourse analysis to compare the way law and science bloggers mark their presence and "position themselves in relation to their arguments and audience" when they disseminate their research. The results point to a functional specialisation and indicate the existence of a contrast between a more personal and subjective style (i.e. more frequent use of first-person pronouns, possessive *my*, object *me* and attitudinal verbs) among scientist bloggers as opposed to a more impersonal style in law blogs (or 'blawgs'), especially when referring to the ideational content of argumentative sections.

Part III "Visibility and Dissemination in Scientific Research Contexts" addresses current research on the practices related to the construction of digital identity and visibility, the emerging conflicts related to public availability and appropriation of scientific culture and the ways of validating and disseminating scientific knowledge in this new context.

In **Chap. 6**, "Predatory Journals: A Potential Threat to the Dissemination of Open Access Knowledge", **Ana Bocanegra-Valle** provides a qualitative analysis of a self-compiled corpus of 50 unsolicited messages sent by predatory publishers. She sets out to identify some 'salient themes' or prototypical features of this type of discourse (e.g. boastful language, promotional self-reference) as well as the accompanying editorial practices commonly used to persuade, or rather, mislead scholars into submitting their work for publication in these journals. Bocanegra-Valle's article reflects on the influence that some institutional and contextual factors that determine scholars' publication decisions may have on the growth of predatory publishing and concludes with a final section on how these malpractices might "flout mainstream academic publication standards", that is, the negative impact they may have on open access, peer-reviewed publications, and hence on the dissemination of reliable, validated research.

Jan Engberg is the author of **Chap. 7**, "Between Infotainment and Citizen Science: Degrees of Intended Non-expert Participation Through Knowledge Communication". He focuses on concepts such as 'popularisation', 'democratisation' and 'dissemination' that point to a specific process of recontextualisation of scientific communication to make it available to a general audience. Engberg's chapter sets out to establish some relevant differentiations among these concepts, which are often used interchangeably, by applying a distinguishing principle: the level of 'participatory ambition' involved between sender (expert) and receiver (layperson) of this information in each case; that is, the experts' varying willingness to create a common space and overcome the existing 'knowledge asymmetries'. To further characterise this continuum, he puts forward a four-stage cline (i.e. *infotainment, dissemination, popularisation* and *citizen science*) which can help readers identify relevant differences in the level of participatory ambition, classify them and explain some resulting changes in the final formulation of a text.

The increasing multimodal "empirical engagement" (Kress, 2015) for description and analysis currently being developed in English for Specific Purposes (ESP) and English for Academic Purposes (EAP) is examined in the following chapters whose aim is to explore the use and combination of different modes and their affordances in digital videos enacting different dissemination practices which perform or fulfil specific purposes. They all share the digital video as medium and the same general communicative purpose: to disseminate scientific content to a diversified, presumably non-expert audience.

Olga Dontcheva-Navratilova explores in **Chap. 8**, "Video Abstracts for Increasing Researcher Visibility", the changes this genre is undergoing with respect to the traditional monomodal character of the written abstract genre, and shows how the multimodal affordances provided by digital communication offer new dissemination opportunities and open up new possibilities for self-presentation and engagement with the audience. Taking a sample of 16 video abstracts created by the researchers themselves in the field of pure mathematics and their corresponding printed versions, she unveils significant changes in their rhetorical structure, with a more descriptive and explanatory character in the former. The identification of the different visual, spoken and written modes and their instantiation patterns in the sample analysed leads her to propose three types of video abstracts, namely, conferential (with a 'personalised' subtype), lecturing and conversational. The different combinations and weight of the verbal and the visual seem to activate specific communicative strategies in each type. A further exploration of the occurrence of metadiscourse markers conveying personality and direct address to the audience in the speech mode reveals the researcher's increasing use of visibility strategies, a growing consideration of viewers as in-group members and contrasting developments depending on the type of video abstract chosen by the researchers. The conversational type, she argues, "represents a radical reformulation", as it appears to be the most interactional type and the one which, according to its view rates, is the most appealing to viewers and consequently has a higher impact on the audience. Dontcheva-Navratilova's study suggests that the recontextualisation process of this migrated genre impinges upon the researcher's efforts not only to persuade the audience to accept the image he/she claims for him/herself, but also to facilitate knowledge comprehension.

The next four chapters also expand the language-only object of analysis, showing a distinctive turn in EAP research to include other modes of communication to explain recontextualisation processes. A clear-cut example is the study carried out by **Noelia Ruiz-Madrid** and **Julia Valeiras-Jurado** in **Chap. 9,** "Reconceptualization of Genre(s) in Scholarly and Scientific Digital Practices: A Look at Multimodal Online Genres for the Dissemination of Science". These authors take the popular science online video as their object of study to examine how scientific knowledge is made available, understandable and attractive to a general audience. Thus, their study delves into how Open Science is professionally constructed to result in what has been referred to as a blending of science, education and entertainment (see Engberg, this volume), a process of recontextualisation that develops a number of specific communicative strategies using an array of different modes. The authors focus their analysis on four professionally produced videos dealing with topics related to Anthropology and Physics from the PBS network, available on YouTube, and apply an inductive and corpus-driven approach first, in order to identify the different modes at work, with the aid of MMAV software, a multimodal annotation tool, then to describe the strategies used, grouped into three functions or purposes, namely, to tailor information, to engage the audience and to build credibility. As a third step, the authors look into the multimodal realisations of these strategies. An interesting observation made by the authors—who consider these videos as "successful" outcomes in bringing science closer to the public—is their 'dynamic' nature, as there seems to be a permanent effort to avoid static images. Although these videos generally make use of a narrator, with embodied modes aimed at engaging the viewer, they always enrich the screen by adding visual effects of different kinds that move or change on the screen; even written text—which is, considering the video format, less frequently employed than images—exploits typography as a mode, with letters being animated to elicit specific feelings and thus contributing to fulfil a determined strategy. All in all, Ruiz-Madrid and Valeiras-Jurado open up a challenging investigative path in search for the recurrent multimodal ensembles and their sequences in these videos, an orchestration of modes that turns a hard scientific question into a story that is both educational and entertaining.

The chapters in **Part IV**, **"Engaging the Audience Through Science Bites"**, examine the means by which researchers exploit their communication skills to reach non-expert viewers, brevity being a determining factor in engaging the audience. We know that one of the main consequences of digitalisation is the access to unlimited information. This has led to a new generation of Internet consumers that want information to be concise, clear and short. We describe the professional practices addressing this search for brevity as 'science bites', evoking the soundbite, a term that originated in North America's news editing rooms in the 1960s to refer to "an audiovisual snippet of media content in which a person is concurrently seen and heard speaking" (Bas & Grabe, 2015, p. 1) without interruption; The three chapters analyse applications of the well-known Three-Minute rule (any presentation should not exceed three minutes to grab the audience's attention span), a defining feature that has been adopted in a new series of scientific communication practices whose main purpose is to disseminate scientific knowledge in the most effective, attractive and concise way, and which has obvious generic relations (as a precedent genre) with the genre of pitch, an intrinsically persuasive professional practice in business communication (Pinvidic, 2019).

Vicent Beltrán-Palanques in **Chap. 10**, "Three-Minute Thesis Presentations: Engaging the Audience Through Multimodal Resources", adopts a multimodal pedagogical approach to the genre of the Three-Minute Thesis presentations. He explains how it can be instructed and implemented, taking into account that doctoral students—at the same time learners and researchers in this very first stage of their academic careers—need, on the one hand, to reformulate content and, on the other, make an effective use of multimodal communication skills, that is, speech and embodied modes, in order to reach non-expert audiences. In this context of situation, the genre stands in very close relationship with the metagenre (Giltrow, 2002) elaborated by the author and aimed at instructing doctoral students on how to produce these presentations, considering all their multimodal affordances.

In **Chap. 11**, "Introducing Science to the Public in 3-Minute Talks: Verbal and Non-verbal Engagement Strategies", the same time constraints apply to the genre chosen by **Juan C. Palmer-Silveira** and **Miguel F. Ruiz-Garrido** for their study, the FameLab scientific talks international competition. In this case, the genre's competitive character and format appear to have evolved as a variant of a genre from the same medium, the TV talent show, and also relies on a metagenre, produced by the event's organisers, which establishes the rules for its realisation, with strict instructions for contestants. The authors chose the ten best young scientists who reached the final event in its broadcast 2020 edition, in which, due to the COVID pandemic, the talks were not delivered to a live audience. Instead, they were pre-recorded with a single stationary camera and uploaded. The authors describe eight engagement strategies, seven of them realised in the spoken mode, and one, termed 'visual impact', in which they include body language and the use of 'external resources', such as props, background and garments. The identification and description of the verbal and non-verbal strategies as the necessary rhetorical tools for an effective dissemination of science in this situated context is an unavoidable first step in any analytical approach to scientific communication in multimodal environments and platforms, as well as in the teaching and training of these communicative skills in the educational context.

Elizabeth Rowley-Jolivet and **Shirley Carter-Thomas** also take in **Chap. 12**, "Research Visibility and Speaker Ethos: A Comparative Study of Researcher Identity in 3MT Presentations and Research Group Videos", the Three-Minute Thesis presentation (3MT) by doctoral students as object of study to compare it with Research Group Videos (RGVs) produced by researchers in university laboratories. The authors collected a corpus of 60 videos, 30 for each genre, to explore the construction of researcher identity at different stages in the academic career as well as from different disciplinary areas, science, technology, engineering, maths and medicine (STEMM) and social sciences and humanities (SSH), taking into account the main modes at work. Modes in 3MTs were restricted to spoken language and embodied modes (vocal effects, posture or movement, gesture, gaze and facial expression), while RGVs added filmic modes (such as stills, moving images, sound effects, animations and graphics), which were incorporated by the professional agency in charge of the postproduction and editing process. The verbal analysis focuses on interactional resources, using AntConc 3.4.0 for quantitative distribution of linguistic features. Below the umbrella function of disseminating scientific knowledge to a general audience, in both genres, researchers use non-technical arguments to explain their work. However, the obvious imbalance of the modal affordances provided by each genre goes hand in hand with their specific purposes and their different contexts of situation and culture. The entertaining function of 3MTs and their limited modal resources make novice researchers "put on a stage performance", maximising embodied modes, particularly gesture, and very colloquial language, while the institutional character of RGVs and their unlimited range of multimodal resources allow for a richer account of identities of senior researchers (institutional, professional and personal) and ultimately a projection of the scientific prestige of their universities.

The last section, **Part V, "Scientific Digital Communication for Research Dissemination: What Lies Ahead?"** has been conceived as an epilogue; **Chap. 13**, "Challenges and Future Directions in Digitally Mediated Research Publication and Dissemination", works as a final summary, where **Vijay K. Bhatia** revisits some of the current issues presented in the volume and provides a personal view of the radical changes in the ways of communicating science online. Bhatia exposes the negative side effects that researchers need to be aware of to maintain the integrity of scientific research, particularly in Open Science. Bhatia argues that the visibility resulting from speedy publication and dissemination appears to be more valuable than "scientifically robust data" and "sustainable results". The author convincingly asserts that the academic community is lowering its guard in terms of research standards, which is leading academia not only to poor quality research but also to ethically questionable practices, such as unverified preprints, thus contaminating scholarly values, particularly in medical and healthcare research. Bhatia appeals to social media to make them aware of their responsibility for the "spread of web-mediated disinformation and misinformation about science". As positive side effects of digital communication in the international context, Bhatia refers to collaborative research and participatory interdisciplinary culture. Bhatia concludes by pointing out the desirable future directions that researchers, editors and digital platforms at large need to follow to turn the current situation around.

All in all, *Digital Scientific Communication: Identity and Visibility in Research Dissemination* aspires to achieve three overriding purposes. Firstly, it has been our intention to gather information about current investigations on scholarly practices unfolded in the digital medium. A particular focus has been placed on the construction of researchers' digital identity and on the visibility of their work, their disciplinary background and their professional concerns. Secondly, we have stressed the importance of the emerging conflicts arising in various research scenarios related to online science communication. Implications are drawn as regards the public availability of information as well as the appropriation of scientific culture in digital settings. Thirdly, we have tried to foreground those avenues of communicating science to the general public through digital practices. Thus, an emphasis has been laid on the dissemination and validation of scientific information online, reflecting upon diverse scholarly and professional practices that aim at knowledge construction and audience engagement in the growingly relevant role played by digital communication.

This volume is directed at postgraduates, doctoral students, practitioners and researchers in the fields of discourse analysis, sociolinguistics, digital media, multimodality and communication studies. We hope that, in sum, this collaborative work makes a valuable contribution to current research on professional practices related to the construction of digital identity and visibility, encourages reflection on the challenges related to the public availability of scientific knowledge, and provides fresh insights into the emerging ways of validating and disseminating scientific knowledge in the digital context.

References

Adami, E. (2015). What's in a click? A social semiotic framework for the multimodal analysis of website interactivity. *Visual Communication, 14*(2), 133–153. https://doi.org/10.1177/1470357214565583

Askehave, I., & Nielsen, A. E. (2005). Digital genres: A challenge to traditional genre theory. *Information Technology & People, 18*(2), 120–141. https://doi.org/10.1108/09593840510601504

Bartling, S., & Friesike, S. (Eds.). (2014). *Opening science: The evolving guide on how the Internet is changing research, collaboration and scholarly publishing*. Springer Open.

Bas, O., & Grabe, M. A. (2015). Sound bite. In G. Mazzoleni (Ed.), *The international encyclopedia of political communication* (pp. 1–6). Wiley. https://doi.org/10.1002/9781118541555.wbiepc063

Bhatia, V. K., & Evangelisti, P., Eds. (2012). *Discourse and identity in the professions: Legal, corporate and institutional citizenship*. Peter Lang. https://doi.org/10.3726/978-3-0351-0321-2

Borgman, C. L. (2007). *Scholarship in the digital age: Information, infrastructure, and the Internet*. MIT Press. https://doi.org/10.7551/mitpress/7434.001.0001

Bou-Franch, P., & Garcés-Conejos Blitvich, P. (Eds.). (2019). *Analyzing digital discourse*. Palgrave Macmillan. https://doi.org/10.1007/978-3-319-92663-6

Bruns, A. (2007). *Produsage: Toward a broader framework of user-led content creation*. Queensland University of Technology. https://doi.org/10.1145/1254960.1254975

Bucchi, M., & Trench, B. (2016). Science communication and science in society: A conceptual review in ten keywords. *Tecnoscienza: Italian Journal of Science & Technology, 7*(2), 151–168.

Calsamiglia, H. (2003). Popularization discourse. *Discourse Studies, 5*(2), 139–146. https://doi.org/10.1177/14614456030050002307

Davies, S. R., & Horst, M. (2016). *Science communication, culture, identity and citizenship*. Palgrave Macmillan. https://doi.org/10.1057/978-1-137-50366-4

Ditlevsen, M. G. (2011). Towards a methodological framework for knowledge communication. In M. Petersen & J. Engberg (Eds.), *Current issues in specialized communication* (pp. 187–208). Peter Lang.

Engberg, J. (2016). Conceptualising corporate criminal liability: Legal linguistics and the combination of descriptive lenses. In G. Tessuto, V. K. Bhatia, G. Garzone, R. Salvi, & C. Williams (Eds.), *Constructing legal discourses and social practices: Issues and perspectives* (pp. 28–56). Cambridge Scholars.

Flowerdew, J., & Wang, S. H. (2015). Identity in academic discourse. *Annual Review of Applied Linguistics, 35*, 81–99. https://doi.org/10.1017/S026719051400021X

Giltrow, J. (2002). Meta-genre. In R. Coe, L. Lingard, & T. Teslenko (Eds.), *The rhetoric and ideology of genre* (pp. 187–205). Hampton Press.

Goffman, E. (1959). *The presentation of self in everyday life*. Doubleday.

Halliday, M. A. K., & Matthiessen, C. (2014). *An introduction to functional grammar*. Routledge. https://doi.org/10.4324/9780203783771

Ito, M., Baumer, S., Bittanti, M., Cody, R., Herr-Stephenson, B., Horst, H. A., & Tripp, L. (2010). *Hanging out, messing around and geeking out*. MIT Press. https://doi.org/10.7551/mitpress/8402.001.0001

Jewitt, C., Adami, E., Archer, A., Björkvall, A., & Lim, F. V. (2021). Editorial. *Multimodality & Society, 1*(1), 3–7. https://doi.org/10.1177/2634979521992902

Johansson, M., Tanskanen, S. K., & Chovanec, J. (2021). *Analyzing digital discourses: Between convergence and controversy*. Palgrave Macmillan. https://doi.org/10.1007/978-3-030-84602-2

Jones, R. H., & Hafner, C. A. (2012). *Understanding digital literacies: A practical introduction*. Routledge. https://doi.org/10.4324/9780203095317

Jones, R. H., Chik, A., & Hafner, C. A. (2015). *Discourse and digital practices: Doing discourse analysis in the digital age*. Routledge. https://doi.org/10.4324/9781315726465

Kastberg, P. (2010). Knowledge communication—Formative ideas and research impetus. *Programmatic Perspectives, 2*(1), 59–71.

Kessler, S. H., Schäfer, M. S., Johann, D., & Rauhut, H. (2022). Mapping mental models of science communication: How academics in Germany, Austria and Switzerland understand and practice science communication. *Public Understanding of Science, 31*(6), 711–731. https://doi.org/10.1177/09636625211065743

Kress, G. (2015). Applied linguistics and a social semiotic account of multimodality. *AILA Review, 28*, 49–71. https://doi.org/10.1075/aila.28.03kre

Lorés, R. (2020). Science on the web: The exploration of European research websites of energy-related projects as digital genres for the promotion of values. *Discourse, Context & Media, 35*, 100389. https://doi.org/10.1016/j.dcm.2020.100389

Lorés, R., & Diani, G. (Eds.). (2021). Disseminating knowledge: Digital discourse in academic and institutional contexts. *Special Issue of European Journal of English Studies, 25*(3), 249–258. https://doi.org/10.1080/13825577.2021.1988262

Lorés-Sanz, R., & Herrando-Rodrigo, I. (2020). New concepts, different approaches: Tackling e-visibility in research project websites. *Revista de Lingüística y Lenguas Aplicadas, 15*, 83–98. https://doi.org/10.4995/rlyla.2020.12782

Luzón, M. J., & Pérez-Llantada, C. (Eds.). (2019). *Science communication on the Internet*. John Benjamins. https://doi.org/10.1075/pbns.308

Marwick, A. E., & boyd, d. (2011). I tweet honestly, I tweet passionately: Twitter users, context collapse, and the imagined audience. *New Media & Society, 13*(2), 114–133. https://doi.org/10.1177/1461444810365313

Mur-Dueñas, P., & Lorés, R. (Eds.). (2022). *Scientific and parascientific communication*. MDPI. https://doi.org/10.3390/books978-3-0365-3822-8

Myers, G. (2003). Discourse studies of scientific popularization: Questioning the boundaries. *Discourse Studies, 5*(2), 265–279. https://doi.org/10.1177/14614456030050020062

Page, R. (2012). The linguistics of self-branding and micro-celebrity in Twitter: The role of hashtags. *Discourse & Communication, 6*(2), 181–201. https://doi.org/10.1177/1750481312437441

Page, R. (2016). Moving between the big and the small. Identity and interaction in digital contexts. In A. Georgakopoulou & T. Spilioti (Eds.), *The Routledge handbook of language and digital communication* (pp. 403–407). Routledge.

Papacharissi, Z. (2011). *A networked self.* Routledge. https://doi.org/10.4324/9780203876527

Pascual, D., & Mur-Dueñas, P. (2022). Dialogic interaction with diversified audiences in Twitter for research dissemination purposes. *Círculo de Lingüística Aplicada a la Comunicación, 90*, 61–79. https://doi.org/10.5209/clac.81307

Pinvidic, B. (2019). *The 3-minute rule: Say less to get more from any pitch or presentation*. Portfolio.

Puschmann, C. (2015). A digital mob in the ivory tower? Context collapse in scholarly communication online. In M. Bondi, S. Cacchiani, & D. Mazzi (Eds.), *Discourse in and through the media: Recontextualizing and reconceptualizing expert discourse* (pp. 22–45). Cambridge Scholars Publishing.

Semino, E. (2008). *Metaphor in discourse.* Cambridge University Press. https://doi.org/10.1017/CBO9780511816802.015

Thurlow, C., & Mroczek, K. (Eds.). (2011). *Digital discourse: Language in the new media.* Oxford University Press. https://doi.org/10.1093/acprof:oso/9780199795437.001.0001

Zerbe, M. J., Cutrufello, G., Hanganu-Bresch, C., & Maci, S. M. (2022). Introduction. In C. Hanganu-Bresch, M. J. Zerbe, G. Cutrufello, & S. M. Maci (Eds.), *The Routledge handbook of scientific communication* (pp. 1–12). Routledge. https://doi.org/10.4324/9781003043782-1

Zhang, Y. (2019). Scifotainment: Evolving multimodal engagement in online science news. In C. Sancho-Guinda (Ed.), *Engagement in professional genres* (pp. 243–258). John Benjamins. https://doi.org/10.1075/pbns.301.13zha

Part II

Scientific Discourse and Professional Practices

2

"Not One of Our Experts." Knowledge Claims and Group Affiliations in Online Discussions of the COVID-19 Vaccine

Ruth Breeze

1 Introduction

One source of concern for public authorities today is the rapid spread of mis- and disinformation on the Internet. Although rumours, propaganda and other information disorders have always been present, perhaps particularly at times of crisis, the Internet and social media now allow messages of all kinds to spread more quickly than ever before. This unprecedented boom in opportunities to disseminate messages that could potentially be inaccurate or harmful became a major cause for concern during the COVID-19 pandemic, when health authorities' attempts to provide guidance on how to prevent infection, for example, were undermined by the circulation of enormous volumes of disinformation (Cairney & Wellstead, 2021; Nguyen & Catalán-Matamoros, 2020; Salaverría et al., 2020).

R. Breeze (✉)
Universidad de Navarra, Pamplona, Spain
e-mail: rbreeze@unav.es

R. Plo-Alastrué, I. Corona (eds.), *Digital Scientific Communication*,
https://doi.org/10.1007/978-3-031-38207-9_2

This phenomenon of user-generated disinformation on a massive scale is a relatively recent issue, and it might not seem entirely clear why it enjoys so much currency. What is the point of joining debates about something one knows nothing about? What impels people to read opinions about important subjects by laypeople who have no specialised knowledge of those issues? These questions have already received some attention in other areas of science that are in the public eye. In the context of reader comments on climate change, for example, Coen et al. (2020) discuss the discursive strategies that commenters adopt to present themselves as having expert status and show how ill-informed users make claims to having valid knowledge that appear to be accepted by a large number of their fellow participants. Users' self-presentation is mainly reinforced by indirect claims to having privileged insights (i.e. through allusions to specific knowledgeable status or first-hand experience) and by direct or indirect reference to technical information, statistics or common sense. In the specific context of COVID-19 vaccines, Scannell et al. (2021) identify a number of strategies used to justify lay contributions to scientific debates in social media, including celebrity endorsements, humour, personal anecdotes, and the appeal to emotions such as fear or hope. They conclude that participants' acceptance of posts presented using such strategies—sometimes over and above genuine expert information—has important implications for public engagement with science-related issues.

At the same time, some previous research has drawn attention to the somewhat surprising way that genuine experts are often rejected or derided on social media. In general, when we lack knowledge about a particular scientific issue, it is commonly assumed that we turn to people who know more, that is, to experts. Indeed, reference to experts has been identified as one of the main ways in which the media boost the authority and credibility of their reporting on scientific issues (van Leeuwen, 2007). However, recent research on social media has challenged our understanding of the public perception of expert status. In particular, investigation of the public understanding of expertise in controversial scientific areas such as climate change or COVID-19 prevention has shown that people are often highly sceptical of media reports drawing on expert statements and that they give full vent to this rejection in online forums (Breeze,

2021; Coen et al., 2020). When interacting online, people seem particularly ready to contest the authority of such sources and surprisingly swift to bring up alternative "expert" figures or lay claim to "expert" knowledge themselves, often on the scantiest of pretexts. Although this phenomenon might sometimes be accounted for in terms of the release from inhibitions resulting from the anonymity of the cybersphere (Suler, 2004), it has also been linked to wider issues. Such positions have been associated with a populist-style rejection of experts as members of an untrustworthy elite (Breeze, 2021) or connected with what Nichols (2017, p. 7) terms the "death of expertise," a phenomenon of "epistemological levelling" caused mainly by equal access to social media, so that average citizens come to believe that their own ideas and opinions about a particular topic are just as valid as those of scientists or doctors with specialised knowledge.

In parallel to this, it has also become abundantly clear that interactions in social media, including discussion forums, are driven more by group affiliations than the desire to exchange information. The individual identity as a self-proclaimed "authorised knower" is subsumed into a group identity, which may be expressed explicitly or implicitly, or remain unstated. As Milan (2015) stresses, the underlying design of social media amplifies the shared, interactive dimension of collective identity, and encourages people to perform and react to each other in highly visible ways. Groups start to coalesce as discursive alignment occurs through a process of "adequation" (Bucholtz & Hall, 2005, p. 599), in which individuals come to understand themselves as "sufficiently similar for current interactional purposes," and "distinction," in which differences are foregrounded and similarities fade into the background. Although both processes are essentially emergent, these aligned identities may also stabilise or "sediment" over time. So when people interact on discussion boards about a particular topic, the positions they take up, the way they ground their claims, their rejection of others' ideas and the attacks they may make on each other are all bound up with group affiliations (Breeze, 2022). The notions held within the in-group about what constitutes expert knowledge, and the strongly held prejudices against information or opinions coming from the out-group, are likely to condition both the views expressed and the kind of interaction that takes place.

In a previous paper, I investigated the popular debate on the COVID-19 vaccine, showing how expert knowledge is problematised and politicised, and how participants claim and contest the right to speak (Breeze, 2021). That paper illustrated how the scientific establishment and pharmaceutical companies are acknowledged as potential sources of authority, as are institutional actors such as the government or health authorities. However, it also showed that "alternative" science and media sources are recognised as having a voice in the discussion and are sometimes credited with more authority than "official" sources, while first-hand experience, even of a rather tenuous nature, is considered a valuable source of information. In the present chapter, I adopt a discourse analytical perspective on reader comments, focusing on individual and group identities and the interplay between the two in order to scrutinise how commenters construct themselves or others as experts, or contest other participants' right to speak.

Critical discourse approaches offer language-oriented analytical techniques for identifying patterns of meaning within a set of texts, in which both micro and macro levels of discourse analysis are integrated (Fairclough, 1992). The choice of lexis, grammatical construction or metaphor observed on the micro level indexes ideological and/or emotional stances, which can be related to broader social representations and their ideological underpinning on the macro level. In particular, the representation of individuals or groups within society, or "social actors" (van Leeuwen, 1996, p. 32), can be analysed to obtain insights into attitudes towards particular entities, which in turn invoke specific ideological positions, particularly when they recur with a certain frequency. Such representations may be made in a deliberate, considered way, or may simply arise in everyday language use or in the course of interaction with other people. For example, use of specific vocabulary to refer to experts (i.e. "armchair experts") can be related to a particular attitude towards experts and accordingly towards the propositions attributed to them, which will allow us to understand better the commenter's view on the issue under discussion (e.g. that "experts" lack appropriate knowledge and should not be relied upon in this case). This can, in turn, be further linked to broader ideological positions that shape the way people's worldview is configured (e.g. the prioritisation of "down-to-earth" values and "hands-on experience" coupled with a rejection of "intellectuals" as "out-of-touch" or "living in an ivory tower"). Such positions are compatible with epistemological assumptions whereby specialised knowledge is rejected and even mocked, in a process of epistemological flattening whereby any participant's words are accorded equal status to any other participant's words, regardless of knowledge, status or experience (Salaverría et al., 2020). This perception is interesting, because it sheds light on why the media's reliance on "experts" in times of crisis, grounded on the assumption that experts are valued and accorded high status, often fails to convince a large sector of the audience.

From the researcher's perspective, without entering the debate about what is actually true or false in a given context, we can thus look at a set of discursive interactions and identify the strategies used to assert the right to speak, demonstrate one's competence, and claim or deny expert status, and we can observe the role of group identification in these. An investigation of this complex area will give us badly needed insights into the nature of public debate on scientific issues, and afford glimpses into how broader social representations of scientific issues are generated and sustained.

The research questions addressed here were:

1. How do commenters ground their authority to make claims about science?
2. How are group affiliations mobilised and what role do they play in such debates?

2 Corpus and Methodology

This chapter is based on a corpus of 10,880 comments published in the UK right-wing popular newspaper *Mail Online* in response to articles about the ongoing development of COVID-19 vaccines, covering the five-month period from February to July 2020 while the first COVID-19 vaccines were undergoing clinical trials in humans. Only articles which received more than ten comments were included (see Breeze (2021) for a full list of the articles used, with headlines, dates and number of comments). The comments were scraped from the online newspaper site using a web scraping application developed within the R statistical computing and text processing environment. They were then cleaned and uploaded to Sketch Engine for lexical processing (Kilgarriff et al., 2014).

Searches were conducted for the lemmas "vaccine" (3304 occurrences), "vaccinate" (204) and "vaccination" (249). The concordance lines containing them were expanded to include the surrounding words (50 words each way). These were then read and re-read in a recursive process in order to establish strategies used to assert the status of the commenter and his/her authority to talk about the vaccine, on the one hand, and to identify allusions to in-group identity and to other groups. Although it is not easy to quantify the number of commenters or comments on each side of the debate, readers can consult data published elsewhere (Breeze, 2021) showing that terms related to expertise were more strongly associated with anti-vaccine comments, while references to knowledge and ways of knowing were more evenly balanced between the two sides.

3 How Do Commenters Ground Their Authority to Make Claims About Science?

Participants who wish to give an opinion about a subject like vaccination, which might be supposed to require specialised expertise, experience in healthcare, or, at least, a good scientific knowledge base, show a certain tendency to ground their views in a knowledgeable self-representation. Evidence from this dataset suggests that this grounding is often explicit, that is, the commenter's "right to speak" is explained or at least alluded to. It may be achieved through alignment with other actors or sources that might be generally accepted as knowledgeable. The grounding may also be implicit, such as when technical language is used that signals the user's command of scientific matters. In what follows, the main strategies identified in the comments are outlined, with an example in each case.

3.1 Explicit Grounding

Participants often explicitly claim expert status when commenting. For example, the (ostensible) professional status mentioned in example (1) would indeed seem to qualify the user to utter an opinion about the vaccine:

(1) I'm actually one of the students working on the vaccine at Oxford... so far everything seems positive!

Some participants mention that they are research scientists, doctors or similar, thus using what Coen et al. (2020, p. 405) call "category entitlements," a frequent interactional resource that people draw on to claim that their status accords them privileged knowledge (Potter, 1996). Users also sometimes simply mention things they have done or habitually do (see 2), which engage with the reader's assumption that someone who has done these things is likely to know about vaccination. In this case, we find a number of participants who allude to "category bound activities" (Coen et al., 2020, p. 405) with the communicative purpose of underscoring their right to speak.

(2) I know, I work in the NHS.

Alignment with named experts, illustrated in (3), is another well-documented strategy that provides a widely accepted source of legitimation (Schmid-Petri, 2017).

(3) Sorry but I'd rather listen to Dr Fauci who said the vaccine will be ready sometime next year

Along similar lines, participants explicitly state their alignment with recognised authorities, in a process by which emphasis on the credibility of the source also boosts the commenter's own opinion (4).

(4) This is the one and only Oxford if they say they all have a vaccine by fall, I will take them for their word.

Identification with Oxford or "UK science" is frequent among the mainly UK-based readers of the *Mail Online*, and such appeals can be seen as invoking in-group loyalty and solidarity, as well as a sense of superiority and national pride. These positive alignments contrast with comments that index rejection of specific individuals or countries, often on ideological or xenophobic grounds:

(5) Not sure I'd want a vaccine made in China!

We also find a broad, general alignment with "science" in general as an entity with authoritative status. In (6), science is held up as the source of expertise, backed by a statement of its relevance to the present case ("the science that created vaccines"), and a suggestion that other people (presumably anti-vaxxers) were aligning themselves against "science" while benefiting from its advances.

(6) If you're not going to trust the science that created vaccines, you don't deserve to be saved by the science that created ventilators.

The vaccine opponents who voice their views on this forum also cite experts, but more usually refer to "common sense" as a source of authority for commenting on vaccine development (7).

(7) Vaccines take 10–15 years to be developed and deemed safe for humans so common sense should tell you that they have had this vaccine for many years.

In many such cases, "common sense" is understood to outweigh the ideas put forward by scientific authorities. In tune with the prioritising of "common sense" over mainstream science, a pattern emerges in which terms such as "scientists" or "experts" are actually used to discredit or devalue a particular position. If we look at the grounding given for rejecting assertions about the vaccine, it is striking that the term "experts" is itself strongly contested in this corpus. Inspection of 208 instances of the lemma "expert" showed that in 175, the term was used negatively and in 25 it was used positively, while in 8 cases it involved statements such as

"I'm not an expert" or "he/she is not an expert." "Expert" often appears in inverted commas (49 cases) and is preceded by denigratory terms such as "armchair," "so-called," "self-proclaimed," "liberal," "social media science," "unqualified" and "deluded." Example (8) represents an ironising use of the word that is rather typical in this corpus.

(8) All aboard the Titanic, the experts say the vaccine is safe.

The phenomenon of vague reference to knowledgeable groups has become so widespread in social media that some authors regard such "non-specific authority references," that is, "experts" or "doctors," as "rumour cues" that should alert us to the possible presence of disinformation (First Draft, 2021). Importantly, however, in the present corpus, most references to experts conveyed misalignment from them rather than uncritical reliance on them (9).

(9) The "doctors" making this vaccine should try it first, if they don't that would tell you a lot.

3.2 Implicit Grounding

Many participants do not explicitly state why they feel they have the right to an opinion on the vaccine. However, it is clear that by using expert or technical language or a specifically scientific tone, many posters implicitly assert their knowledge of science and biomedicine. Here the reader shows their familiarity with the vaccine development process (10) and implicitly aligns with the scientific establishment:

(10) Phase three is probably the most important phase as it tests the new vaccine on a greater number of people. 70–90 per cent of drugs that enter Phase III successfully complete.

In example (10), the adoption of an expert voice involves the use of technical terms ("phase three"). But importantly, it also employs "truth statements," that is, statements that are formulated in such a way as to leave no space for doubt (e.g. "70–90 per cent of drugs that enter Phase

III successfully complete"). Gilbert and Mulkay (1984) identified such statements as part of the empiricist repertoire, which gives the impression that the comments "follow unproblematically and inescapably from the empirical characteristics of an impersonal natural world" (p. 56). The use of conventional scientific–academic discourse plays a role in creating this effect: thus (11) uses the academic formula "it should be remembered that" to introduce a "truth statement."

(11) For all the hype about a vaccine for Covid-19, it should be remembered that 15 years after its close cousin, SARS, hit the world, there is still no vaccine for it.

Such statements are far from being the prerogative of the "pro-vaccine" commenters. In fact, both here and elsewhere they are used by both sides in the vaccine debate (Breeze, 2021). In (12) the authoritative tone achieved through academic terms such as "historically" is combined with scientific-sounding phrases such as "on a natural decline" to argue that the vaccine is no longer needed because the virulence of the disease is already waning, a statement that is clearly false.

(12) Historically, vaccines arrive on the scene each time a disease is on a natural decline. By the time each is introduced, the number of cases is falling away & this will be the case with this one.

One further point about this "scientific" voice is that it is frequently contrasted to what "laypeople" or "most people" think. In (13) the authoritative voice is used to correct what the commenter thinks is a common fallacy.

(13) The majority of people think the difficulty in creating a vaccine is the vaccine itself and that's not the case. It's the reaction from said vaccine that can only be assessed by a prolonged period of testing.

Another approach to claiming the authority to speak is used by those commenters who emphasise first-hand experience. Although both sides of the vaccine debate make use of this approach, it is particularly frequent in the anti-vaccine group. In (14) the speaker ironises that their experience of non-vaccination and no illness "must be a coincidence." Experience is clearly weighted more heavily in this argument than the implied argument in favour of vaccination that is being refuted.

(14) Never had a vaccine in my life never been sick never been to hospital must all be a coincidence.

Similar to this, many comments appeal to "common sense" or what might be termed "naïve logic," without actually naming a source. In (15) the commenter poses a rhetorical question to undermine the notion that a vaccine can be produced.

(15) But they told us yesterday that the virus mutates like crazy. So if it's constantly mutating, how will a vaccine work?

Other examples point to apparent logical inconsistencies that the commenter has noticed, leading to an interpretation of what is happening that points towards conspiracy theories (16). By grounding comments like these in "naïve logic," it is to be supposed that these participants are appealing to like-minded people who are capable of "seeing through" the official line on this subject. The implicature (i.e. that normal intelligent people can spot the inconsistencies in the official discourse) would tend to appeal to other readers with a critical stance and contribute to the construction of solidarity among that group.

(16) It's pretty strange how many coronavirus vaccines have been popping up out of thin air. It's almost like they were ready for this pandemic in regards to pushing their vaccines on us.

Following on from this general distrust of the official line, many well-known conspiracy theories such as 5G (see Nguyen & Catalán-Matamoros, 2020) appear in this comments corpus, and we find these bolstered by generalised suspicion of the establishment, the pharmaceutical industry ("big pharma") and the political elites (Breeze, 2021;

Schmid-Petri, 2017), as in example (17), or of a mysterious conspiracy involving some or all of these, as in (18):

(17) Somebody somewhere is going to get filthy rich when the vaccines are made public.
(18) Their plan is coming together nicely. Somehow I knew any new vaccine they developed for Covid would only be effective for "1 year."

By stirring up suspicion of those who are promoting the vaccine in financial, scientific and political terms, comments of this kind echo populist discourses that place "the people" in counter-position to the "elite" in a way that is profoundly political.

4 How Are Group Affiliations Mobilised and What Role Do They Play in Such Debates?

The above examples have been explained within a framework of individual, ideational grounding that might actually distract us somewhat from the vigorous and even aggressive interaction within the forum, which is driven by group identification as much as by rational argumentation. At least three interesting emotion-driven group dynamics are established: one in which users appeal to others to comply with vaccine rules for the common good, another whereby users construct themselves as potential victims of forcible vaccination and/or as willing to defy the authorities on this, and another in which users' prejudices are mobilised to sway participants towards or away from a particular position.

In (19), we see the argument in favour of vaccination as a social act in order to protect weaker members of society, which surfaces in various shapes and forms and is clearly an appeal to people's solidarity.

(19) Can you both please explain to me a different way vulnerable people can be protected from coronavirus?

However, considerable emotional intensity is conveyed in many defiant posts that anticipate compulsory vaccination (20).

(20) THE GUY WHO TRIES TO PIN ME DOWN AND FORCIBLY INJECT ME WITH A VACCINE IS GONNA WISH HE NEVER TRIED TO PIN ME DOWN AND FORCIBLY INJECT ME WITH A VACCINE

It is also interesting that the vaccine is constructed as dangerous by many of the participants who are critical towards it, yet those who accept it are derided as cowardly (21).

(21) Only scared sheep will take this so-called vaccine.

At the same time, insulting language is used to refer to those who are willing to comply with the authorities, generally insinuating that those people are unintelligent (22) or mindlessly conformist (23).

(22) Can't wait to see the queue of simpletons just begging to get injected.
(23) Those two sheeple are so brainwashed there's no point trying to explain.

Emotions associated with the assertion of national group belonging on the one hand, and xenophobia on the other, are also frequently mobilised in comments such as (24–26).

(24) He's not one of our experts. We have a far more talented pool of academia at Oxford and Cambridge.
(25) They stopped giving this [the TB vaccination] to the British children to give to the immigrants coming in... disgusting.
(26) Who in their right mind would want to get a vaccine invented and produced by China?!

In general, both the appeals to the in-group and the attacks on the out-group identified here are imbued with emotional force, but the emotions involved are different depending on the commenter's attitude towards vaccination. Pro-vaccine commenters tend to appeal to people's sense of responsibility or duty, framing vaccination altruistically as a way to protect vulnerable people (as in 19). Vaccine-critical participants are more likely to convey anger and aggression (20), ridicule (21), contempt (22, 26), and hostile or contemptuous feelings towards other groups (25, 26), although such feelings are not exclusively found on this side (e.g. 24 invokes in-group national identity within the pro-vaccine camp, as well as exclusion of "other" groups). Humour and irony seem to be more frequent among those opposed to the vaccine, perhaps as a way of building solidarity among like-minded people by mocking those who think otherwise. Fear is present on both sides of the debate, but is framed rather differently: fear of COVID-19 and other diseases in the case of pro-vaccine participants, and fear of vaccination (either of the vaccine itself or of force exerted by the state) among anti-vaxxers. On both sides, the appeal to emotions is likely to reinforce in-group solidarity and encourage discursive group alignment by building positive feelings towards those who hold similar opinions, focusing negative energy on out-group members, and giving expression to shared fears.

5 Discussion

Although this study lacks a quantitative orientation, the overview it provides of the strategies used to ground comments and interact with other users brings to light certain patterns concerning the social representation of what constitutes reliable knowledge. The discussion itself is highly polarised, but it is extremely interesting that the participants not only disagree with each other but actually underscore the validity of their contribution in two contrasting ways: (1) by invoking experts in the field or adopting the subject and style of communication attributed to experts or (2) by challenging this epistemology grounded in respect for expertise and proposing instead an understanding in which "common sense" and personal experience take priority. These two main approaches to grounding knowledge strongly resemble those found in other contemporary studies of reader comments on controversial scientific issues analysed elsewhere (Breeze, 2021; Coen et al., 2020; Schmid-Petri, 2017; Vilar-Lluch, 2022). Against this background, three aspects warrant further discussion, namely, the driving force of group affiliation in social media, the erosion of epistemological hierarchies in the anti-vaccine group and the significance of these findings for scientific communication practices.

First, it is interesting to consider the strategies underlying the intragroup and intergroup dynamics in the processes we have observed. In the interactions observed here, participants persistently fell into two groups, namely, pro- and anti-vaxxers. In their interactions, they both attacked the other group and encouraged members of the in-group through various community-building strategies. This can be compared with the findings of other papers on scientific controversy in social media. Researchers working on Twitter noted that "pro- and anti-vaccination users formed two polarized networks that hardly interacted with each other" (Milani et al., 2020, p. 364). In particular, they found that anti-vaxxers formed a group with strong cohesion fostered by frequent emotional support as well as regular retweeting, whereas pro-vaccination tweeting was more fragmented, with looser connections. The analysis of interaction in the present corpus also suggests that the social minority group (anti-vaxxers) worked harder to put its views across and interact with other users, as well as giving emphatic support to in-group members. They made prominent use of explicit grounding (using the first person, expressing feelings, referring to concrete groups and places), and often used strong language to denigrate vaccine supporters. Interestingly, humour and irony were mainly found in the anti-vaxxer tweets, which echoes the findings of some previous research (Scannell et al., 2021). Moreover, opponents of the vaccine presented it as dangerous, ineffective or unnecessary, and ulterior motives were attributed to its promoters, often with allusions to conspiracy theories. As Jaspal and Yampolsky (2011) noted, in such situations of conceptual conflict, the stakeholders that would conventionally be regarded as less powerful (here, the anti-vaxxers) often make a concerted attempt "to 'upgrade' their polemic social representations to hegemonic level" by supplying elaborate theories and raising the emotional tone. By contrast, the pro-vaccine group, which was here more aligned with official policy, tended to use implicit grounding, and to relay information from official sources. Its contributions were thus less personal and less colourful. Perhaps because this group reflected the mainstream opinion, there were fewer expressions of strong feelings, less emotional bonding, and more implicitly grounded arguments. Overall, this group did not succeed in mobilising a strong sense of solidarity. In terms of the consequences of this difference, it is possible that some users might find the anti-vaxxer posts more original and exciting (as it exploits the impetus of the "protest group"), and the solidarity among this group more attractive, than the duller, less stimulating message and tone used by the pro-vaccine community. We should also note that since the present study focuses on reader comments rather than tweets, the dynamics observed here is somewhat different. Unlike the tweeters in Milani et al. (2020), the pro- and anti-vaccine groups encountered here did interact, on occasion rather aggressively, a phenomenon also observed in other studies of controversy in reader forums (Breeze, 2022; Corona, 2017). We can surmise that this could be an artefact of the medium itself: the forum pushes readers to see each other's comments, so that they are more likely to confront users who hold different views, whereas the algorithms underlying some other social media expose users only to opinions that coincide with their own, in the so-called echo chamber effect (Terren & Borge-Bravo, 2021).

Second, moving on from the strategies used to examine what these interactions reveal about the representation of expert knowledge, if we take a step back to consider the wider context, we will see that the erosion of knowledge hierarchies observable in the anti-vaccine comments here offers strong similarities to the "epistemological flattening" that has been described as one of the consequences of social media (Salaverría et al., 2020), and which some have interpreted as one of the symptoms of a "regime of truth change" (Harsin, 2015). Coen et al. (2020) assert that the type of argumentation found in social media discussions of scientific issues "indicates that the very idea of expertise is a fluid concept that is not captured satisfactorily by 'traditional' definitions" (p. 413). In this, it is perhaps productive here to follow the line developed by Nichols (2017), who ventured further into the underlying reasons for this apparent rejection of experts. Three specific reasons he discusses are (1) the pseudo-democratising nature of the Internet and social media (in which all players seemingly enjoy equal status); (2) the "customer" model in higher education, which undermines not only the status of teachers, but also that of specialised knowledge itself; and (3) trends in the media whereby news and information are increasingly trivialised and presented as entertainment. Cumulatively, these trends might make people less likely to accept the idea of expert knowledge, and believe that everyone's opinion is equally valid, culminating in the growing distrust of expert systems predicted by Giddens (1994). Paradoxically, Nichols (2017) suggests, the trend towards increasing democracy and ever wider dissemination of information has not led to the rise of an educated and well-informed public. Rather, it has generated a generation of ill-informed and resentful citizens who distrust the notion of intellectual achievement, with obvious undertones of political populism, in which "the people" and their ways of thinking, doing and knowing are placed in counter-position to those of untrustworthy "elites." Of course, it may well be that, as Carvalho has discussed (2007, p. 223), the decision about what constitutes scientific news and who should define what is true in science is ideologically determined. Similarly, as Cairney and Wellstead (2021) indicate, trust in government clearly conditions attitudes and behaviour in different countries: low accountability meets with high distrust. Nonetheless, it is crucial that we find ways of communicating the complex issues arising out of scientific research, because as Nichols (2017) says, when ordinary citizens believe that no one knows more than anyone else, democratic institutions themselves are in danger of falling either to populism or to technocracy—or in the worst case, a combination of both. One interesting detail here is that the explicitly grounded anti-expert messages hold a very specific kind of appeal, engaging popular audiences in a way that a highly technical message simply cannot. We can compare these to the pro-vaccine comments analysed in the present study, whose messages are constructed around classic appeals to serious scientific knowledge, yet come across as dull and lacking in conviction.

Third, following on from this, we should ask what these findings might mean for scientific communication practices. As people are increasingly likely to seek scientific and medical information online (Anesa & Fage-Butler, 2015), health authorities need to rethink the way they communicate with different sectors of the population. Since we have seen that group affiliations are particularly important in shaping how people receive and relay messages, communication specialists need to consider how these loyalties might be leveraged to ensure a positive impact. At the same time, it is clear that education has a major role, since it is undeniable that enhanced scientific literacy and media literacy could play a role in combatting information disorders in this area (Vraga et al., 2020). As Kienhues et al. (2020) suggest, people need to be encouraged to grapple with complexity in order to build informed trust in scientific disciplines and engage more actively with their epistemological challenges. However, as Nguyen and Catalán-Matamoros (2020) point out, to tackle information disorders concerning the online reporting of scientific issues, it is not enough merely to provide fact-checking services or refute disinformation. Instead, those responsible need to gain a deeper understanding of how digital media "facilitate or hinder the interaction between rational factual knowledge on one hand and emotions, values and beliefs on the other" (p. 325), and how this "shapes public engagement with the health and science issues at stake" (p. 325). In this sense, it is not clear whether health authorities should adapt their communications policies to incorporate some of the styles favoured in social media (i.e. by presenting serious information in a more popular, affective way in order to engage with audiences on different levels), or whether they should combat the emotionalisation and trivialisation of serious scientific issues through educational campaigns designed to raise awareness among different publics. In either case, more discourse-based studies are needed to examine the dynamics of such forums, the mechanisms of polarisation, and the crucial role of group affiliations in such interactions.

References

Anesa, P., & Fage-Butler, A. (2015). Popularizing biomedical information on an online health forum. *Revista Ibérica, 29*, 105–128.

Breeze, R. (2021). Claiming credibility in online comments: Popular debate surrounding the COVID-19 vaccine. *Publications, 9*(3), 34. https://doi.org/10.3390/publications9030034

Breeze, R. (2022). Performing right-wing political identities on reader comments pages. *Pragmatics & Society, 13*(1), 85–106.

Bucholtz, M., & Hall, K. (2005). Identity and interaction: A sociocultural linguistic approach. *Discourse Studies, 7*(4-5), 585–614. https://doi.org/10.1177/1461445605054407

Cairney, P., & Wellstead, A. (2021). COVID-19: Effective policymaking depends on trust in experts, politicians, and the public. *Policy Design and Practice, 4*, 1–14. https://doi.org/10.1080/25741292.2020.1837466

Carvalho, A. (2007). Ideological cultures and media discourses on scientific knowledge: re-reading news on climate change. *Public Understanding of Science, 16*(2), 223–243.

Coen, S., Meredith, J., Woods, R., & Fernandez, A. (2020). Talk like an expert: The construction of expertise in news comments concerning climate change. *Public Understanding of Science, 30*(4), 400–416. https://journals.sagepub.com/doi/full/10.1177/0963662520981729

Corona, I. (2017). A life well lived of a lady well loved. In R. Breeze & I. Olza (Eds.), *Evaluation in media discourse. European perspectives* (pp. 179–202). Peter Lang. https://doi.org/10.3726/b10531

Fairclough, N. (1992). Linguistic and intertextual analysis within discourse analysis. *Discourse & Society, 3*(2), 193–217.

First Draft. (2021). Finding misinformation with "Rumour Cues". 25 February. Retrieved June 22, 2022, from https://medium.com/1st-draft/

Giddens, A. (1994). Risk, trust, reflexivity. In U. Beck, A. Giddens, & S. Lash (Eds.), *Reflexive modernization: Politics, tradition and aesthetics in the modern social order* (pp. 194–197). Polity Press.

Gilbert, N., & Mulkay, M. (1984). *Opening Pandora's box: A sociological analysis of scientists' discourse.* Cambridge University Press.

Harsin, J. (2015). Regimes of posttruth, postpolitics, and attention economies. *Communication, Culture and Critique, 8*(2), 327–333.

Jaspal, R., & Yampolsky, M. (2011). Social representations of the Holocaust and Jewish Israeli identity construction: Insights from identity process theory. *Social Identities, 17*(2), 201–224.

Kienhues, D., Jucks, R., & Bromme, R. (2020). Sealing the gateways for post-truthism: Re-establishing the epistemic authority of science. *Educational Psychology, 55*, 144–154.

Kilgarriff, A., Baisa, V., Bušta, J., Jakubíček, M., Kovář, V., Michelfeit, J., Rychlý, P., & Suchomel, V. (2014). The Sketch Engine: Ten years on. *Lexicography, 1*, 7–36.

Milan, S. (2015). From social movements to cloud protesting: The evolution of collective identity. *Information, Communication & Society, 18*, 887–900. https://doi.org/10.1080/1369118X.2015.1043135

Milani, E., Weitkamp, E., & Webb, P. (2020). The visual vaccine debate on Twitter: A social network analysis. *Media and Communication, 8*(2), 364–375.

Nguyen, A., & Catalán-Matamoros, D. (2020). Digital mis/disinformation and public engagement with health and science controversies: Fresh perspectives from Covid-19. *Media and Communication, 8*(2), 323–328.

Nichols, T. (2017). *The death of expertise: The campaign against established knowledge and why it matters*. Oxford University Press.

Potter, J. (1996). *Representing reality: Discourse, rhetoric and social construction*. SAGE.

Salaverría, R., Buslón, N., López-Pan, F., León, B., López-Goñi, I., & Erviti, M. C. (2020). Desinformación en tiempos de pandemia: Tipología de los bulos sobre la COVID-19. *Profesional de la Información, 29*, e290315.

Scannell, D., Desens, L., Guadagno, M., Tra, Y., Acker, E., Sheridan, K., Rosner, M., Mathieu, J., & Fulk, M. (2021). COVID-19 vaccine discourse on Twitter. *Journal of Health Communication, 26*(7), 443–459.

Schmid-Petri, H. (2017). Politicization of science: How climate change skeptics use experts and scientific evidence in their online communication. *Climatic Change, 145*, 523–537.

Suler, J. (2004). The online disinhibition effect. *Cyberpsychology & Behavior, 7*(3), 321–326.

Terren, L., & Borge-Bravo, R. (2021). Echo chambers on social media. *Review of Communication Research, 9*, 99–118. Retrieved June 22, 2022, from https://rcommunicationr.org/index.php/rcr/article/view/94

Van Leeuwen, T. (1996). The representation of social actors. In C. R. Caldas-Coulthard & M. Coulthard (Eds.), *Text and practices: Readings in critical discourse analysis* (pp. 32–70). Routledge.

Van Leeuwen, T. (2007). Legitimation in discourse and communication. *Discourse & Communication, 1*(1), 91–112.

Vilar-Lluch, S. (2022). Social reaction to a new health threat: The perception of the Covid-19 health crisis by British and Spanish readerships. In A. Musolff, R. Breeze, K. Kondo, & S. Vilar-Lluch (Eds.), *Pandemic and crisis discourse: Communicating Covid-19 and public health strategy* (pp. 185–206). Bloomsbury.

Vraga, E. K., Tully, M., & Bode, L. (2020). Empowering users to respond to misinformation about Covid-19. *Media and Communication, 8*(2), 475–479.

3

Utmost Hybridity: Promotional Trends in Technology Disclosures

Carmen Sancho-Guinda

1 Introduction

Can the loss or absence of identity become an identity feature? What if, in actual fact, a given genre has never had a distinctive identity of its own but an appropriated format and discourse, hybridity being its only way to acquire distinctiveness? This seems to be the case of some technology disclosures, also known as 'university technologies', 'university inventions', 'technology solutions', 'innovations', 'technology descriptions', or 'marketing abstracts', a genre still in the making present in the websites of every university and indicative of the institution's prestige, based in turn on its research dynamism, versatility and productivity, and, in sum, of its service to society. French historian of science and epistemologist Gaston Bachelard (1981/1934) notes the inextricable relationship between science and technology: the latter refers to the procedures and resources

C. Sancho-Guinda (✉)
Universidad Politécnica de Madrid, Madrid, Spain
e-mail: carmen.sguinda@upm.es

R. Plo-Alastrué, I. Corona (eds.), *Digital Scientific Communication*,
https://doi.org/10.1007/978-3-031-38207-9_3

used by the former and to the ability to experiment with or implement them. Technology can be then understood as the 'praxis' of science (as its research outcome) and scientific phenomena as 'constructs' simultaneously theoretical and (potentially) practical. Bachelard's (1970) neologism 'phenomenotechnique' aptly captures such dual nature.

With the Genre Analysis framework (Swales, 1990, 2004; Bhatia, 1993, 2004) as backdrop and point of departure, this chapter will explore how hybridity configures the discursive identity of technology disclosures (henceforth TDs) at the Universidad Politécnica de Madrid (hereafter UPM) at three different spheres of communication: *co-textual*, *semiotic*, and *strictly verbal*, as well as point to the challenges and dangers these phenomenotechnical accounts pose for effective dissemination and promotion. Although words are signs and therefore also semiotic, I will deal with verbal language as a separate meaning-making category for clarity's sake.

2 Impact of Digitalisation on the Dissemination of Science and Technology: The Case of Technology Disclosures

Together with interconnectivity and partly caused by it, hybridity is a direct consequence of digitalisation. A closer contact (or even contiguity) among textual genres by means of hyperlinks and website buttons has brought about a 'contagion' of rhetorical, visual, multimodal, and linguistic traits, (multi)purposes (e.g. simultaneously informative, promotional, and entertaining), and interinfluences of varied sorts (e.g. communal, environmental, and, as said, multimodal). The affordances of computer-mediated communication have certainly made interactions more accessible and immediate (today dissemination is indeed rapid, mostly thanks to the emergence of increasingly briefer online genres) and have shortened the distances between communities of practice (Wenger, 1998), now more heterogeneous, permeable, and transient. Digital affordances have also related formerly distant genres and discourses that originally belonged to quite dissimilar realms (technical, legal, journalistic, fictional, and, in general, leisure-oriented) and even blended some of them, such as the 'scifotainment' detected in sci-tech news by Zhang and O'Halloran (2014) and Zhang (2019).

Among discourse researchers, analysable phenomena around digitalisation are, in addition to interconnectivity, hybridity, interinfluence, permeability, multimodality, and immediacy, the recontextualisation of information and communicative purposes and the tensions between situatedness and multilocation and between factual and opinionated interactions (Kuteeva & Mauranen, 2018). These two authors also note a 'pastiche assemblage' of messages that, along with hypertextuality, contributes to connectivity. Other intriguing objects of study within digital discourses and genres are their affective interactivity (Petroni, 2011) and the growing informality that underpins the so-called 'cyberspace college' (Puschmann, 2015), which affect how scholars interact with one another, project their identities, and acquire visibility (Luzón, 2018; Lorés-Sanz & Diani, 2021). Computer-mediated communication (CMC for short) has also changed the text consumption habits of printed dissemination (Mur-Dueñas, 2018) and (re)shaped, adapted, and repurposed traditional dissemination practices. These have turned hybrid and, according to Pérez-Llantada (2016), are open-ended enough so that they can meet the emerging communicative needs of professional communities.

One fact is evident: CMC has thwarted the conventional expectations of academic discourse, impersonal and 'sanitised' over decades (Swales & Feak, 1994; Hyland, 2006, 2009; Biber & Conrad, 2009) and almost deprived of emotivity (Dzung Pho, 2013), explicit appraisal (Hyland, 2006, 2009), and expressions of attitude (Hyland, 2006). These discursive features are gradually being replaced by an emotional marketing and branding (Gómez Cabranes, 2013), a foregrounding of the author, especially in multimodal settings such as the video-essay (Maier & Engberg, 2013), and the already mentioned informal tone of the 'cyberspace college' alluded to by Puschmann (2015), whose boundaries between the public and private are fuzzy and may incorporate a didactic and even 'scifotainment' touch.

A slippery issue arises in relation with CMC genres: Which ones should be truly considered 'digital'? Only those originally generated for online dissemination and/or interaction? Or may we include those that existed in print before the digital era but have somehow adapted to CMC, such as the research article, the abstract, and the editorial online? Seemingly, no consensus seems to have been reached in this matter, as remarked by Kuteeva and Mauranen (2018) and Lorés-Sanz and Diani (2021). In this sense, TDs make a singular case because they for sure existed much longer before their appearance in digital media, but presumably as a quasi-occluded genre institutionally filed and discreetly reaching out to potential sponsors, business partners, and patent examiners through the university's mediation. It is reasonable to suppose that, if TDs exhibited, the format of an average scientific abstract when made visible online for the first time in the early 2000s, their 'occluded shape' would have been the same. Example 1, a 2011 screenshot, shows a typical abstract-like instance from a small corpus I compiled in 2011 during a research stay at the University of Michigan to study the TDs of the Michigan Research Corridor, a research consortium integrated by the University of Michigan, Wayne State University, and Eastern Michigan University. In the sample, we can identify the standard IMRD narrative of science, yet here the introduction merely consists of the research purpose intertwined with the method, and the result is followed by an extremely brief discussion that states the centrality of the investigation, a rhetorical move usually found right at the outset in introductions. There are classificatory labels under the title (identity and file numbers) and on the right-hand side margin (the technological category), which also lodges the name and contact data of the university mediator, some prominent member of staff of the Office of Technology Transfer (OTT from now on) or Venture Centre, and the names of the inventors/researchers, clickable to be traced in their respective departments, but without any contact data. At the bottom of the informative space are a hyperlink leading to the full technology description and the institutional emblem, which visually claims for intellectual property.

Example 1 Wayne State University. Screenshot

Derivation of Molecular Targets of Cancer and Aging Using Cellular Immortalization-Specific Gene-Expression

Case ID:
02-612
Description:

Detailed Description

WSU faculty developed an approach to identify gene-expression changes in spontaneously immortalized human cell lines as compared to non-immortal cells. This approach led to the identification of molecular targets required for the long lifespan spontaneously acquired by cancer cells. These targets may provide a potential means to intervene in the process of aging or cancer development of a series of genes involved.

File Number: 02-612

Disease: Cancer

Direct Link:
http://wayne.technologypublisher.com/technology/5151

Category(s):
Diagnostic
Bookmark this page
Download as PDF
For Information, Contact:
Joan Dunbar
Director of Biotechnology Development
Wayne State University
(313) 577-5542
jcdunbar@med.wayne.edu
Inventors:
Sorin Draghici
Olga Studitskaia
Michael Tainsky

WAYNE STATE UNIVERSITY

Example 2, another screenshot from the same corpus and also belonging to the category of biomedical sciences, displays a structured abstract variant. Its headings divide the technical information more efficiently but at the expense of taking up more space. While the university logo is present, formulaic institutional language is reiterated (my underlines): "The University of Florida...", "University of Florida researchers..." and the tone becomes personal with the use of 'exclusive we' (my emphasis). It is not clear whether the possessive adjective in the expression "Our revolutionary method enables capture of the desirable properties of each component in one product" refers to the institution or to the researchers, nor who the author of the statement is. Surprisingly, no researcher names are given, nor that of the mediator—only a telephone number and an e-mail address are provided. This kind of anonymity was frequent at the time.

At their early monomodal stage, we may conclude, TDs tended to be anonymous or to prevent contact with the scientists adopted the format of the average scientific abstract, either structured or conventional (yet focusing more on the product than on the research process, unlike the research article abstract), were subjected to a high institutional control and written in a formal style, more often than not impersonally, occasionally showed overt promotional features within the technical discourse, such as adjectives of the type 'novel', 'innovative', or 'revolutionary' (all of them in Example 2), were monological (intertextuality and interdiscursivity being rare), and appeared isolated under the corresponding website button, frequently labelled as 'Research and innovation', 'University technologies', and so forth. Their connectivity, therefore, was rather low.

Example 2 University of Florida

Amino Acid Terminated Polymers for Graft Copolymerizations

Background

The University of Florida is seeking companies interested in commercializing a novel technique using amino acid terminated polymers to produce graft copolymers. It is often desirable to combine two or more polymers during manufacturing to yield a product that exhibits properties from each polymer. However, several problems, including an inconsistent blend, are encountered with current graft copolymerization methods. University of Florida researchers have improved the grafting of copolymers by eliminating many of the problems associated with current techniques.

Applications

Methods and compositions for enabling the preparation of homogeneous mixtures of polymers which exhibit the properties of each polymer.

Advantages

- Enables the mixing of polymers heretofore not compatible.
- Provides polymer alloys exhibiting combinations of properties heretofore unavailable.

Technology

University of Florida researchers have developed an innovative graft polymerization method that uses amino acid terminated polymers to graft copolymerize two or more polymer systems. Our revolutionary method enables capture of the desirable properties of each component in one product. This technology uses free-radical chain transfer methods to graft copolymers, eliminating the need for an energy source such as radiation. The use of amino acid for chain transfer offers an improvement over other graft copolymerization approaches, and provides homogeneous grafted polymers.

To discuss this technology with a licensing officer call (352)392-8929 or email jmuir@ufl.edu and ask about record UF ID: 1687

UF UNIVERSITY of FLORIDA

Some decades later, they would become inter-semiotic (i.e. in the main bimodal, but perhaps multimodal, depending on the institution's budget and projection policies), visibly authored, multifunctional, and more personal, informal, heteroglossic, and networked, as will be explained in the coming sections to describe the three-level hybridity that defines the genre in my own university.

3 Co(n)textual Hybridity: Genre Constellations, Blends, and Bends

Nowadays TDs are being inserted into evermore intricate netnographic ecosystems, but between the visible genres for dissemination (e.g. research articles, abstracts and highlights, newsletters, press releases, online science news, and TDs themselves) and those for commercialisation or marketing (e.g. online patent applications, filed patent documents in the patent and trademark offices of each country or of transnational areas such as the EU, and once again TDs within an institutional scope) lie some occluded genres, invisible to both the higher education community and the general public. OTT's staff reveal that before a technology is disclosed in the institutional website, scientists must have summarised their invention or discovery in written form, presented it orally to some committee in charge of its evaluation, and undergone one or more interviews. Feedback on the technology is normally supplied in a screening report filled out by OTT workers.

All of these intermediate genres, however, are hardly accessible by us applied linguists and remain to be analysed as to their standards formats and rating criteria, as well as to the correlation between success, language and semiotic uses, their extension/duration, and the rhetorical conventions in each of them. It may be a matter of little time that some voices are raised to vindicate more transparency in the evaluation of university technologies, as is happening with the reviews of journal articles.

For the time being, nonetheless, all we can witness is the online constellation (Bhatia, 2004; Swales, 2004) of four genre colonies (Fig. 3.1): those for *technology production*, for *dissemination*, for the *commercialisation or marketing*, and for the *training* of the educational community in the culture of innovation. Well-known and important technology-production genres are the local technology and innovation contests in which mixed teams of students, teachers, and personnel can participate,

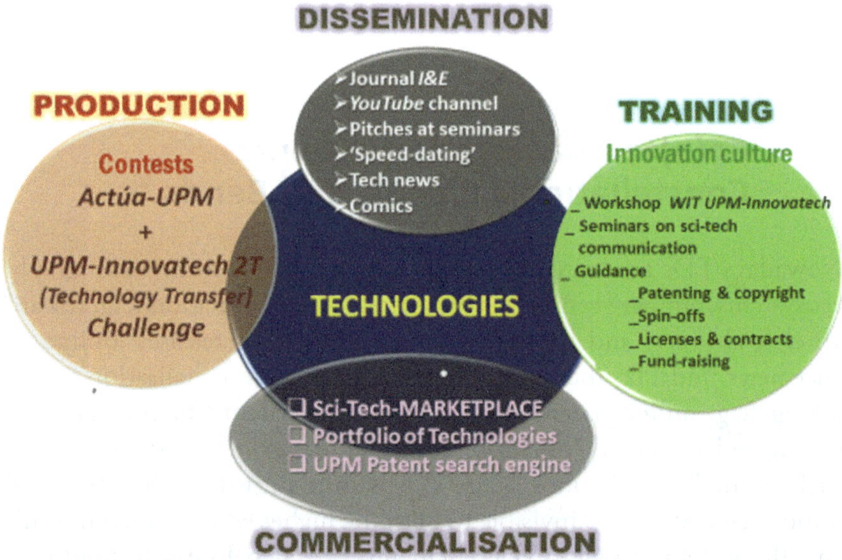

Fig. 3.1 Genre colonies in the online constellation/ecosystem at the UPM

such as 'Actúa-UPM' and 'UPM-Innovatech'. The dissemination colony includes the in-house journal *I&E*, technology-related UPM videos uploaded in the YouTube channel, elevator pitches at R+D institutional seminars, 'speed dating' events, technological news in the university portal, and comic books that narrate the experience of technology creation to high-schoolers and young undergraduates. The colony for commercialisation or marketing includes filed and granted patents, trackable with the aid of a university search engine, and the institution's TDs, accessible with exactly the same format in two different online locations alike: the university's technological 'marketplace' and its technology catalogues/portfolios (Centre for Support of Technological Innovation-UPM, 2018) in Spanish and English. The training colony comprises workshops and seminars for preparing contestants and permanent guidance on patent and copyright issues, spin-offs, the procurement of licences and contracts, and fund-raising strategies.

Thus, the network of TDs spreads out in multiple directions. Any TD may result from training programmes and in-house technology contests, be the topic of any of the online genres of the dissemination colony, and accessed through two different website paths devoted to commercialisation—the university's marketplace and its technology portfolios. Ultimately, TDs may give rise to spin-offs that customarily transfer the university's research results or can be patented, in which case they are then permanently kept in the university archives with links to the national and international patent and trademark offices where they are registered.

In light of all this, the TD behaves as an 'amphibian genre' that traverses all four colonies and also various intra- and interdisciplinary fields and institutional departments, sections, and activities, little by little becoming more dialogic and multimodal. Who could have predicted these networked ramifications in its beginnings as an unobtrusive and practically isolated text with a borrowed format? Its meteoric evolution concerning form and content, at least in my own university, begs the key question of what is to be considered 'prototypical', 'bent', and 'innovative' among higher education institutions around the world. Clearly, TDs at the UPM do blend purposes, audiences, semiotic resources, and genre features and discourses. They exemplify the 'context collapse' (Marwick & Boyd, 2011) characteristic of digitalisation, initially associated with social media and obscuring the identification of a single audience profile. The voices behind TDs from the UPM address a variety of audiences and communities of practice (experts in the field, lay people, and experts from other technical fields more or less akin or tangential) and play disparate roles such as those of researcher, instructor, disseminator or science journalist, institutional ambassador, and even entertainer, the latter especially in titling practices, sometimes tinged with humour. This role blend is equally manifest in recent scholarly genres online such as the graphical and video-abstract, having been prompted by the new and more dialogic, engaged, and participative paradigms of science dissemination (Stocklmayer, 2013), in which the mediating figure of the science journalist is slowly disappearing and researchers are consequently supposed to undertake the mediation between their expert communities and audiences with varying degrees of knowledge.

If corporate branding boils down to "creating differentiation and preference" (Mautner, 2010a, 2010b, pp. 141–142), then it is logical to expect wide variations in format and style across institutions, as long as vital technical information is covered in the TD description. Going global through digital dissemination will, paradoxically and unavoidably, bring to the surface plenty of localisms and diversity, so prototypicality cannot be but rhetorical, unless the OTT of some university devises a TD model that really makes a difference regarding benefits and therefore it becomes widely imitated and exploited. Counter to this remote possibility is the search for institutional distinctiveness, because following Hyland's (2012) concept, we may consider TDs 'representational genres' just like corporate and institutional websites. As for bending, it is also improbable on an individual plane since private intentions are curbed by the institution, although this may have vested interests itself: for example, it can obtain profit via contracts for transference of research results through the creation of spin-offs for technologies that would not yield economic benefits if they did not reach the market (OTRI Universidad de Granada, 2022).

Finally, we may wonder whether or not the TDs under study constitute innovation. Convention and innovation go hand in hand: the former is indispensable for the latter to occur (Tardy, 2016, p. 171), but the TD is still in its infancy as a genre, without established, socially accepted conventions (rhetorical, stylistic, semiotic) defining prototypicality and distinguishing innovative instances that depart from the norm. There is an additional criterion for innovation signalled by Tardy (2016, pp. 9–10)—the perception of the genre as 'effective' or 'successful' by its intended audience. This complicates matters further because, as said, audience profiles are blurry and only prudential observation can ascertain efficacy and success. What community should we focus on and how much time must pass to tell? And what type of 'success' do we mean? Monetary income? International prestige reflected by the number of students registered due to the cutting-edge aura of the institution?

To recapitulate, context and cotext exert a reciprocal influence: social environments generate certain types of texts/genres and texts/genres in turn build and reflect social realities. In search of institutional distinctiveness and increased citizen participation in the dissemination of technologies and their evaluation, future TDs might become multimodal and furnish opportunities, synchronous or asynchronous, to make comments and record lay impressions and expert feedback, just as many journal article reviews are becoming open-access and fostering collective involvement. Only time will tell.

4 Semiotic Hybridity: The Triangle of Mimesis, Pertinence, and Ornament

From a semiotic standpoint, there are two remarkable features displayed by the TDs issued by the UPM: *mimesis* and a *fluctuating functional use* of visuals. Format mimesis, intentional or not, can be taken as visual interdiscourse; in this case, as the presence of academia in the technology description, given the strong optical resemblance of TDs with the current institutional format of academic posters, as can be perceived by contrasting Figs. 3.2 and 3.3. Both genres contain an introductory caption with the title, which in TDs is enriched with two slight expansions: one in the form of a subtitle or micro-summary, and the other as an obligatory abstract called 'briefing'. Contrastingly, abstracts are redundant in posters (and thus often omitted) because the whole of them serves its synthesising role. Whereas the rest of captions may vary their number and distribution freely in the scientific abstract, UPM's TDs infallibly use a second final caption for the contact data of scientists/inventors and OTT personnel mediators. It could be argued that the overall mimetic 'camouflage' of TDs as academic posters seeks to enhance the technological proposal as more rigorous or 'serious' than any other visual format might suggest, thus helping promote the technology further.

Curiously, then, optical scientific mimesis becomes a promotional strategy, but precisely through the concealment or disguise of overt promotionalism, although OTT personnel from the UPM has not as yet confirmed any intentionality underlying their visual choices, most probably made intuitively. Another feature that meets the eye is the insertion of journalistic-like content highlights as a deference to the reader, in order to mark the salience of certain information items visually. These highlights may or may not go between scare quotes, which personalise the description as if they were stressing an excerpt of interview with the

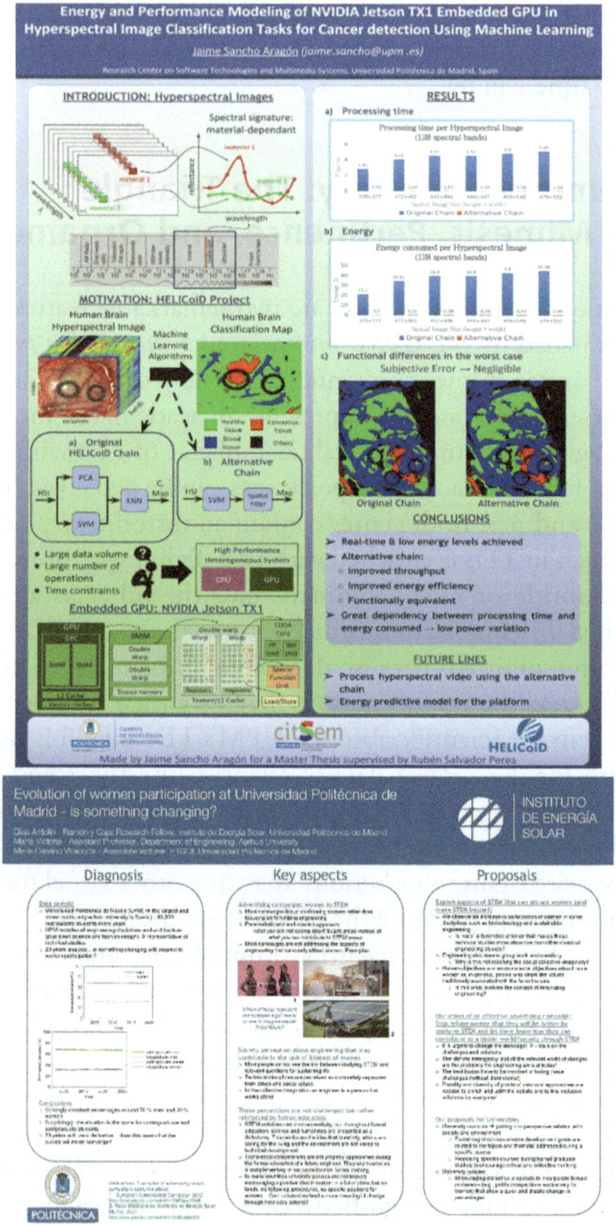

Fig. 3.2 Two contemporary samples of thesis poster formats at the UPM from the masters (above) and PhD (below) programmes

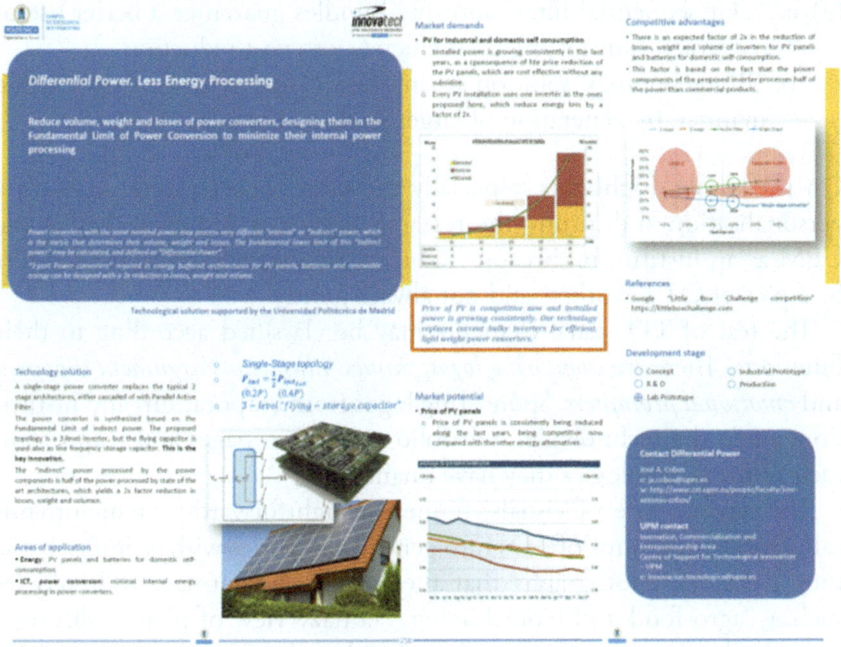

Fig. 3.3 Average TD sample from the UPM's technologies portfolio (English version, 2018). (Reproduction authorised)

researchers (imitating singled out interviewee soundbites), or the text excerpt had been selected by the interviewees themselves. These highlights are written in a bigger italicised font size and in a different colour (light blue). In Fig. 3.3, I have taken the liberty to mark the highlight with a frame so that it can be spotted more easily.

Promotion is present in every highlight, but it may appear *in isolation*, reminiscent of slogan advertising (Examples 3a and 3b), a formula hardly present in patent applications because patents take for granted the qualities of novelty and usefulness, or in addition *define* the technology or *encapsulate* its basics; for example, with the specific features that make the invention or discovery advantageous, as shown in Examples 4a and 4b.

(3) a. "Simple, practical and effective"
    b. "High productivity, non-polluting agriculture"

(4) a. "Our sequential filters and ring bundles guarantee a better use of resources, transferring their efficiency to cost reduction"
b. "Biotechnology solution to increase plant yield in a sustainable manner, by generation of larger seeds and plants"

On the whole, highlights, especially the ones between quotes, increase personalisation and 'usurp' the researchers' voices and even that of the UPM as an institution, because TD texts are written and translated just by a pair of OTT workers and not always jointly.

The rest of TD visual elements may be classified according to their functions. There are *sponsoring logos*, *frames*, *illustrators/argument supports*, and *emotional prompters*. Sponsoring logos from corporations and institutions open the technologies' portfolio across two pages and reappear in each of the technologies they have financed.

The second type of visuals, frames, straightforwardly or metonymically, sorts out groups of TDs into general categories with a similar basis: they are stock-photographs that thematically agglutinate technologies such as 'agro-food and biotechnology' (a hazy view of plant stalks on a green background, one with a dangling dew drop), 'energy & environment' (a row of offshore windmills in a yellowish dusk or sunrise atmosphere), 'health' (a web of luminous neuronal connections), 'IT, Internet & mobile' (a bunch of optical fibres), 'Transport, security & space' (a shot of traffic in a modern city area lit by night), and 'Construction and infrastructure' (a tall building reflected on a skyscraper's window pane). These six images are displayed twice: small-sized and all together as a visual table of contents, and separately and blown up to introduce each category, preceding the verbalised sub-table of contents within each of the technological categories.

Argument supports or illustrators are adjacent to the verbal TD text, either intercalated or as columns on the margins, and consist in photographs, assorted graphs, tables, or diagrams. They may illustrate what the technology looks like or how it works, prove the efficacy or convenience of the technical features described, or evince the social knowledge gaps and service lacks that justify the technology in question. However, they are not always mentioned in the verbal text but presented as complementary independent information. The fourth and last kind of visuals, emotional prompters, are also interspersed in the TD's verbal text, but their pertinence to the technical topic is tangential or even questionable. For instance, a thriving green sprout, a smiley African kid, or a pair of hands scooping seeds may appear in given agro-food TDs; emojis and abstract well-being images of children and elders, or the metaphorical depiction of an autistic boy isolated behind a glass wall enrich some health TDs; a drawing of a human head with visible cogwheel gears inside and photographs of open books are found in IT TDs; the earth seen from the outer space, earthly landscapes and interacting people in TDs having to do with transport, security, and space; and shattered glass or rubble in construction and infrastructure TDs. The purpose of these prompting pictures is at the same time ornamental and emotional: they evoke a positive emotional response in readers, so as to make them realise the need for the technology or appreciate its beneficial outcomes, and sprinkle a note of lightness, colour, and a decorative reportage-quality to the technical description.

Semiotic hybridity, we may infer, operates at different levels and combines varied strategies. At a pre-text introductory level, it frames the TD inventory of the portfolio/catalogue with the idiosyncratic presence of numerous corporate logos, and through stock images organises the general table of contents in a metadiscursive fashion. In the TD proper, at a compositional level, it unfolds a mimetic academic-poster layout that imbues the document with a scientific air while mingling diverse types of visuals with different functions. Granted, we cannot assure (and neither can the OTT personnel itself) that every strategy is conscious, but there is no questioning that their final effects definitely reinforce promotion.

5 Verbal Hybridity: Whose 'Babel' Is It?

As advanced at the beginning of this chapter, my reason for a separate study of hybrid features in the written text, labelled as 'verbal hybridity', intends to dissipate ambiguity and gain precision, since the heading 'discursive hybridity' would necessarily encompass visuals. Obviously, words are arbitrary signs and in good logic should be included under the 'semiotic' category, but counterposing verbal and visual elements runs the risk of assuming that visuals are always intentional and 'visible'; that is, optically perceivable and crystal-clear to the addressee, which does not happen systematically with some graphics and subtle collocational and compositional practices. Furthermore, there are elements, such as the 'highlights', with 'one foot in each camp', verbal and visual. By subsuming all visual components under the 'semiotic hybridity' class, I acknowledge their potential as signs or meaning-makers, intentional or not, and differentiate them from the written word, most often deliberate.

The TD genre, unlike the academic poster, is attributed to seasoned researchers and expected to be subjected to an extraordinarily high degree of institutional intervention, as it has become a (quasi) representational text over the last pair of decades and its 'templates' are more strictly regulated than those of the poster with regard to rhetorical moves, overall design, and visual insertions. Actually, the researchers' preferences find little or no room at all. It can be said that, while posters are fundamentally 'idiolectal' because it is the researcher, no matter how novice, who makes rhetorical and compositional decisions within the bounds of the several standard templates approved of or proposed by the university or research centre; TDs are fully 'institutional'. Along with the issue of *authorship*, which I will deal with later on in this section, other chief differences detectable in their respective verbal texts are the value and placement of reference sources to build a *scholarly ethos* and the control of *interpersonal distance*.

Posters are written in a depersonalised style that may be made up for in its associated 'teaser', 'blitz', or 'poster-madness' presentation, but no other genre inherently complements the TD, whose interpersonality fluctuates under the institution's dictates and the preferences of the actual writer, seldom the scholar's. Admittedly, other constellated genres may portray complementary facets of the social milieu surrounding the TD, but none, to my judgement, forming a genre chain as close and immediate as the dyad poster + presentation. We have already seen that the journalistic-like highlights in UPM TDs might include personal pronouns and even be framed by scare quotes that lead to a fake attribution 'usurping' authorship, and that evaluative vocabulary could be overt and 'idiolectally passionate' (let us think of promotional qualifiers such as 'unique', 'pioneer', or 'exceptional', in Examples 5a, 5b, and 5c, all from 2018—my italics).

(5) a. This represents a *unique* technological solution able to identify early detection for voice pathology, its treatment or even identify a person biometrically. (GLOTTEX®)
b. Compared to conventional systems, the proposed sensor system is a global *pioneer* in measuring the pH in the air in both open and closed environments. (TECHNICAL PROTECTION FOR THE HISTORICAL–CULTURAL HERITAGE)
c. Due to *exceptional* properties such as high flexibility, unique electronic and optical properties, high transparency or ultra fast charge carriers, two-dimensional nanomaterials are suitable for a wide range of applications. This allows our technology to fulfil different markets. (TDTRANSFER. AUTOMATIC TRANSFER FOR LAYERED NANOMATERIALS)

This blend of factual and promotional language concentrates mostly in the abstract or briefing and in the description of the technology (entitled 'technology solution'), both paragraphed, but can also appear in the bulleted sections of 'market demands' and 'competitive advantages' (Table 3.1 shows the rhetorical moves of the TD genre at the UPM). Such overt promotion was practically unthinkable in early TDs online and, if present, was more objective, as can be read on the first line of the Florida University TD from 2011 in Example 2: the adjective 'novel' does, in effect, apply to a literally newly devised technique. The adjective 'unique' (Example 5a), which semantically goes one step further beyond than 'exceptional' (Example 5b) in what concerns originality and exclusivity, has been the object of revealing studies with the aid of Lexical and Corpus Pragmatics and Relevance Theory. Díez-Arroyo (2020, p. 167), for one, notices the existence of 'broadened readings' of the term in which the originally non-gradable form acquires gradability and collocates with suchlike adverbial modifiers as 'completely' and 'truly' in the laudatory descriptions of automobiles. To denote innovation and brand identity, it collocates with common car constituents (e.g. rear lights, (steering) wheels, etc.), design elements (e.g. inlays, patterns, stitching, etc.), and

Table 3.1 Rhetorical moves of the TDs at the UPM and their major features

Rhetorical Move	Headings/ Sections	Features
Summarised introductory description of the technology (various degrees of brevity) + promotion • **Title** (maximum brevity) • **Subtitle** (intermediate brevity) • **Briefing or abstract** (slight expansion)	No headings	Titles and subtitles may be humorous and very promotional Briefings and subtitles may present fluctuating personalisation Contained in a coloured caption
Detailed description of the technology	'Technology solution'	Paragraphed Frequent mixture of complete sentences with fully conjugated verbs and telegraphic style
Claim for research and market 'niches'	• 'Areas of application' • 'Market demands' • 'Market potential' • 'Competitive advantages'	Bulleted Frequent mixture of complete sentences with fully conjugated verbs and 'telegraphic' style May include visuals
Ethos building through research group credentials	• 'References' • 'IPR' (Intellectual Property Rights')	Bulleted Fluctuating 'telegraphic' style Patents and copyrights as a separate heading
Indication of the technology's development stage	'Development stage'	Multiple-choice cells to be ticked off: concept, R&D, lab prototype, industrial prototype, production
Contact with scholar(s) and institution	No heading	Data contained in a coloured caption Separate data from scientists/ inventors and university mediators

with the words 'experience' and 'pleasure' conveying unforgettable hedonistic episodes (e.g. unique onboard/driving experience/pleasure). The term, as Díez-Arroyo gathers, has come to signify 'special' and 'unequalled' and no longer means 'the only one'. Rather than singling out and classifying entities and objects, it has turned into an emotion-raiser.

Recourse to marketing expressions, mixed with the technical register expected from the genre by specialists (and, what is more, combined with institutional formulas and with fluctuations between personalisation and a 'sanitised' style), produce a sort of 'Babelic' interdiscourse that may come across as confusing and hard to reproduce by scientists/inventors, and above all by engineers, accustomed as they are to what Sales (2006) calls 'restrained persuasion'. Her scrutiny of engineering grant proposals and interviews as part of her ethnographic research has brought to light the reluctance of engineers to overtly 'sell' their work by means other than empirical data or mathematical reasoning. This fact positions most TDs in a 'no man's land' between the restrained persuasion of the hard and applied sciences and the purely advertising discourse of promotional genres, as rendered in Fig. 3.4. Of course, it remains to be determined the

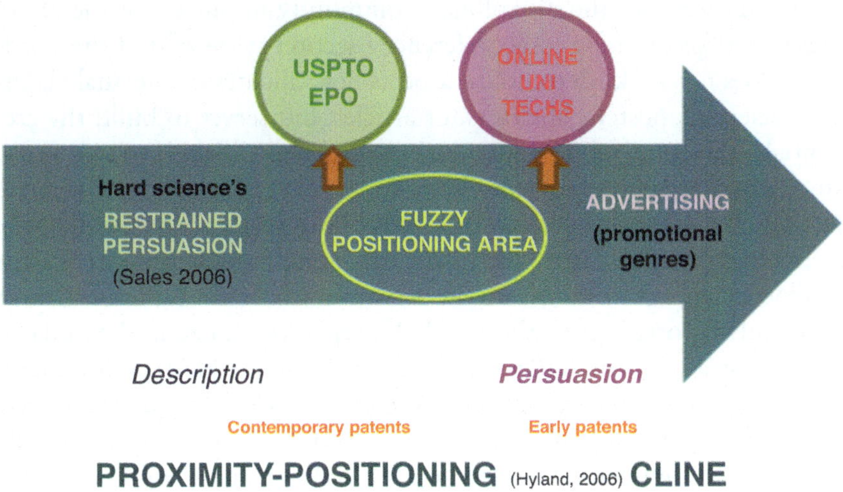

Fig. 3.4 The 'no man's land' of TDs at the UPM compared with contemporary patent texts from the U.S. Patent and Trademark Office (USPTO) and the European Patent Office (EPO)

average percentage of intervention from OTT staff in the adjustment of the scholars' original submission, because the organisation of the information is presumably very different, unless the scientists/inventors have previously been provided with the sequence of headings under which to fit the different technical aspects. OTT writers, who are professional marketeers or have received specialised training on merchandising at the institution, will most likely be responsible for the marketisation of the description. Another unknown is whether the scientists/inventors have the final say-so as to the degree of promotionalism. Are their suggestions and complaints really attended to and are they satisfied with the final TD version? The identity negotiated by the UPM in its TDs is, in Hyland's terms (Hyland, 2005a, 2005b, 2010, 2012), more positioning- than proximity-oriented. Simply stated, it is more inclined to an expert–lay interaction that minds the knowledge deficit of outsiders to the community of practice than to an expert–expert communication that displays signs of exclusive belonging, stints on glosses, and follows the disciplinary conventions only insiders know.

There are rhetorical moves whose sections fulfil a 'bridging' function between experts from the institution and lay or mixed readers, sacrificing 'proximity traits' of the disciplinary community in pursuit of the aforementioned 'positioning'. The 'references' section at the end of the document does not back up technical aspects nor validates intellectual claims as in scientific posters, abstracts, or articles, but serves to build the credentials, the *ethos*, of the individual researcher or of the research team, among which patents and copyrights are so valued that call for a separate section. The references section in TDs is but the direct promotion of the technology creators embedded in the overarching marketisation of their product. Figure 3.5 enumerates in decreasing order the credentials most frequently reported across the 134 TD samples contained in the updated English version of the UPM's *Technologies Portfolio* (2018), still current. Those involving intellectual property, external collaboration, experience, and institutional support outnumber several others and supposedly inspire more confidence from prospective sponsors, partners, and clients. Prestige is unevenly handled, since it yields a relatively high percentage of reporting on an individual basis, but very low as collective credential.

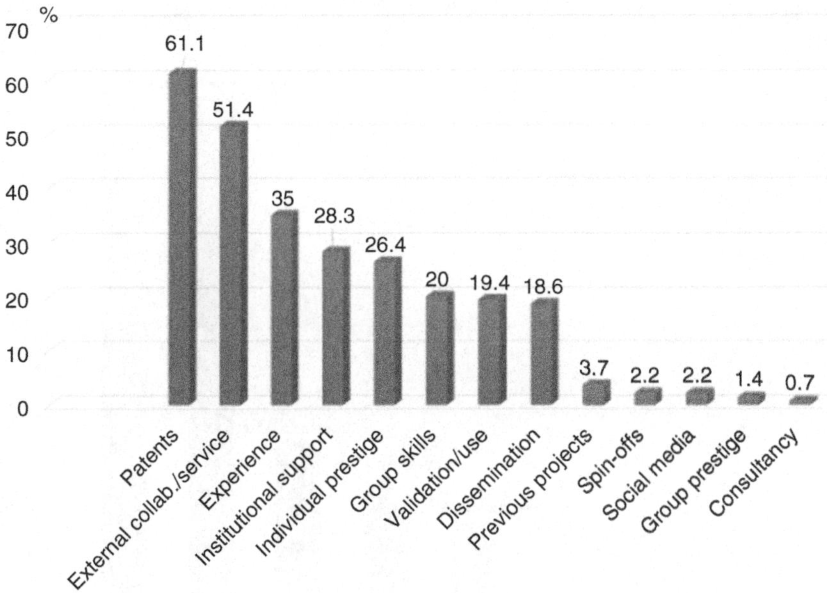

Fig. 3.5 Credentials most frequently reported in the TD's references section at the UPM

Titles may as well function as reducers of interpersonal distance. My analysis of the 134 titles in the *Portfolio* attests to the unstoppable overt marketisation of technological discourse. As can be readily observed in Fig. 3.6, nearly 95% of titles resemble advertising slogans or consist in the technology's commercial name followed by some condensed definition, a salient feature, or a slogan-like sentence or phrase. The trend in all title categories, nevertheless, is to attract and persuade lay readers by approaching the technology to them with familiar expressions and storylines (e.g. idioms), discourses (that of advertising) and intertexts (e.g. famous quotes or cultural references from popular wisdom), to sting their curiosity with paradoxical associations between concepts or meanings, or to achieve nearness by means of explicitness, commonly through instances mentioning a procedure or action with a gerund or through intensifiers/boosters that emphasise an important property or benefit. 'General' titles

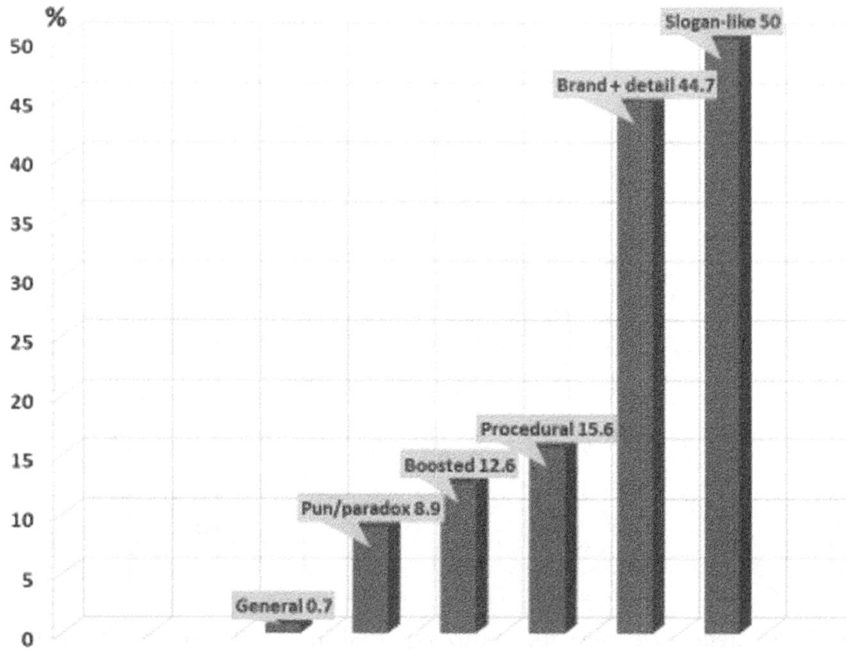

Fig. 3.6 Title typology in the UPM's current English version of its *Portfolio of Technologies*

consisting in a flat phrase or sentence defining the essentials of the technology are very few. Below are Examples (6–10) of each title category. Some may belong to more than one type:

(6) Puns, paradoxes* and cultural intertexts**

a. OOPS! Ontology pitfall scanner!
b. SILFRARED. The dark side of light*
c. DYLAN-Q. Answers are not blowin' anymore**

(7) Slogan-like/Commercial brand + definitory detail*

a. WOODEX. In the simplicity is the answer
b. GLOTTEX. Your voice treats you, your voice identifies you

c. PROMETEO. The robot that saves lives*
d. BULBUICK. Bigger and faster plant growth

(8) Procedural-like

a. IBIS. Ensuring a safe exploitation of space
b. b.YESOGRAF. Improving the properties of gypsum and plasters

(9) Boosted

a. F2TE3. Highest efficiency and freedom in the design of buildings
b. TRACHEO4LIFE. Highest reliability for emergency tracheotomies

(10) General

a. Light multilayer façade system
b. Platform for detection of neurological disorders

After this succinct analysis of the genre co(n)text, of its structure and discursive promotional strategies, Goffman's (1981, p. 128) famous question inevitably surfaces: "Who is speaking?" Whose 'babel' of lay and specialised discourses is this? Authorship poses quite a thorny issue, as TD writing seems to be a collaborative endeavour. In accordance with the Goffmanian inventory of speech roles, it is not difficult to deduce that the university is the 'principal' committed to the meaning transmitted, present in ritualistic phraseology such as "Technology solution supported by the Universidad Politécnica de Madrid" below the caption containing the title, subtitle and briefing, and before the heading of the detailed description; or in formulaic briefing openings such as "Researchers from the UPM have found/propose/present (…)" (Examples 11a and 11b, italics mine). It is not hard either to identify 'figures' (i.e. the actors or characters in the technology story), normally the researchers (Examples 12a and 12b, my emphasis), but distinguishing the 'author' (i.e. the text designer) and the 'animator' (i.e. the actual writer) is not that easy. Who outlines the content and writes it in the end?

(11) a. *Researchers at the Universidad Politécnica de Madrid* have developed and manufactured a new concept of laminated material of recycled conglomerate textile with metallic or polymeric coating. (TEKSTEELA)

b. *The Universidad Politécnica de Madrid, in collaboration with* the Sabadell Hospital, has developed an algorithm that automatically calculates the insulin dose that should be administered to a continuous subcutaneous insulin infusion or 'insulin pump', in order to maintain healthy blood glucose levels. (ARTIFICIAL PANCREAS)

(12) a. In Leuko Labs, *we identified* a crucial medical need: to avoid the serious infections that frequently occur in cancer patients treated with chemotherapy. (LEUKO LABS)

b. We designed a rubber-cement compound with high mechanical strength and we use to make ecological concrete blocks, which reduce the carbon footprint and improve the thermal comfort conditions of buildings. (Neoblock)

Authorship is shared, with variable percentages depending on the field, by the scientist/inventor and the OTT staff, especially the Marketing Coordinator or Commercialisation Manager (at the UPM). Some universities may commission the writing to hired graduate students or interns previously trained, as the Michigan Research Corridor has been doing in the last decades, but always under the permanent supervision of the licensing staff, who proofreads and edits the successive versions of the TD text and has the final say for its publication. Graduates are selected there by their writing skills, undergo a crash course on the writing of marketing abstracts (in 2011 it lasted less than 24 hours), and receive feedback on the abstracts they write over the next three to four weeks. While it may strike as odd that graduate students and interns are preferred over experienced scientists/inventors to disseminate their technologies, OTT staff argue that the latter usually have problems in delimiting what the actual innovation is and are unaware of the persuasive dimension of the TD document, which they conceive of as solely descriptive. At the UPM, TD writing falls under the non-transferable remit of marketing-trained OTT personnel and is accomplished through consultation with inventors and scientists. Goffman's question triggers off the classical interrogative posed by Berkenkotter and Huckin (1995): Who owns the genre, in the end? The institution, because of the document's undeniable representational role in the official website lodging it and given its possible impact on the number and type of prospective students and business partners? The technology creators, without whom absolutely nothing would be possible? The OTT staff, whose action is decisive to make TDs understandable to mixed and lay audiences and attractive to potential sponsors and buyers? Do all enjoy fair recognition?

6 A Closing Thought

Hopefully this chapter has shed some light on the TD genre, still under-researched, and on the role played by its hybrid nature (between dissemination and outreach, individuality and community, the verbal and the visual, the personal and the impersonal, the technical and the promotional, and the academic and professional spheres, being occluded at some stages and openly accessible in others) in the construction of institutional identity. Through a first approach to a young genre with a singular digital format, I have shown a small slice of university culture reflected on a creative constellation of genre colonies and the inner interactional dynamics of higher education communities. As Hyland (2012) cautions, "behind every individual's engagement in a professional existence lies an institutional identity constructed through countless interactions", but it is no less true that behind institutional identities there are individuals who make decisions as to format, content, conventions, and connections. In subsequent research phases, I intend to tackle the narrative focalisation strategies of TDs (Sancho Guinda, 2022 forthcoming), which will plausibly undergo substantial transformations by incorporating extra affordances and increasing the protagonism of technology creators and OTT writers. Much awaits to be researched: from ethnographic responses clarifying institutional and individual motivations and goals, to contrasts between TD corpora across national cultures and university statuses, and reception studies on the TDs' intended audiences, members of the university community, and outsiders. It is my wish that this small contribution paves the way for further intriguing discoveries.

Acknowledgements The author thanks the editors for the opportunity to disseminate her work, as well as for their dedication and immense patience.

References

Bachelard, G. (1970). *Nouméne et microphysique.*

Bachelard, G. (1981/1934). *El nuevo espíritu científico.* Nueva imagen.

Berkenkotter, C., & Huckin, T. N. (1995). *Genre knowledge in disciplinary communication: Cognition/culture/power.* Lawrence Earlbaum..

Bhatia, V. K. (1993). *Analysing genre. Language use in professional settings.* Longman.

Bhatia, V. K. (2004). *Worlds of written discourse. A genre-based view.* Continuum.

Biber, D., & Conrad, S. (2009). *Register, genre and style.* Cambridge University Press.

Centre for Support of Technological Innovation-UPM. (2018). *Portfolio of technologies-Universidad Politécnica de Madrid.* https://www.upm.es/sfs/Rectorado/Vicerrectorado%20de%20Investigacion/Oficina%20de%20Transferencia%20de%20Resultados%20de%20Investigacion%20(OTRI)/Comercializacion/Catalogo_tecnologias_UPM_innovatech_2018%20[EN].pdf

Díez-Arroyo, M. (2020). The adjective "unique": More than a buzzword in the language of advertising. *Ibérica, 40*, 149–170.

Dzung Pho, P. (2013). *Authorial stance in research articles: Examples from Applied Linguistics and Educational Technology.* Palgrave Macmillan.

Goffman, E. (1981). *Forms of talk.* University of Pennsylvania Press.

Gómez Cabranes, L. (2013). Las emociones del internauta. In L. Flamarique & M. D'Oliveira Martins (Eds.), *Emociones y estilos de vida: Radiografía de nuestro Tiempo* (pp. 211–243). Biblioteca Nueva.

Hyland, K. (2005a). *Metadiscourse: Exploring interactions in writing.* Continuum.

Hyland, K. (2005b). Representing readers in writing: Student and expert practices. *Linguistics and Education, 16*(4), 363–377. https://doi.org/10.1016/j.linged.2006.05.002

Hyland, K. (2006). *English for Academic Purposes: An advanced resource book.* Routledge.

Hyland, K. (2009). *Academic discourse: English in a global context.* Continuum.

Hyland, K. (2010). Constructing proximity: Relating to readers in popular and professional science. *Journal of English for Academic Purposes, 9*(2), 116–127. https://doi.org/10.1016/j.jeap.2010.02.003

Hyland, K. (2012). *Disciplinary identities: Individuality and community in academic discourse.* Cambridge University Press.

Kuteeva, M., & Mauranen, A. (2018). Digital academic discourse: Texts and contexts. *Discourse, Context and Media, 24*, 1–7. https://doi.org/10.1016/j.dcm.2018.06.001

Lorés-Sanz, R., & Diani, G. (2021). Disseminating knowledge: The effects of digitalised academic discourse on language, genre and identity. *European Journal of English Studies, 25*(3), 249–258. https://doi.org/10.1080/13825577.2021.1988262

Luzón, M. J. (2018). Constructing academic identities online: Identity performance in research group blogs written by multilingual scholars. *Journal of English for Academic Purposes, 33*, 24–39. https://doi.org/10.1016/j.langcom.2021.02.004

Maier, C. D., & Engberg, J. (2013). Tendencies in the multimodal evolution of narrator's types and types and goals in research genres. In M. Gotti & C. Sancho Guinda (Eds.), *Narratives in academic and professional genres* (pp. 149–173). Peter Lang.

Marwick, A. E., & Boyd, D. (2011). "I tweet honestly, I tweet passionately": Twitter users, context collapse, and the imagined audience. *New Media & Society, 13*(1), 114–133. https://doi.org/10.1177/1461444810365313

Mautner, G. (2010a). *Language and the market society: Critical reflections on discourse and dominance*. Routledge.

Mautner, G. (2010b). Language and communication design in the marketplace. In R. Wodak & V. Koller (Eds.), *Handbook of communication in the public sphere* (pp. 131–154). De Gruyter Mouton.

Mur-Dueñas, P. (2018). Disseminating and constructing academic knowledge in online scholarly journals: An analysis of virtual special issue introductions. *Discourse, Context & Media, 24*, 43–52. https://doi.org/10.1016/j.dcm.2018.04.010

Oficina de Transferencia de Resultados de la Universidad de Granada. Spinoffs Universidad de Granada. *Spin off: Qué son y para qué sirven*. Retrieved September 27, 2022, from https://spinoff.ugr.es/cms/menu/info-otri/spin-off-que-son-y-para-que-sirven/

Pérez-Llantada, C. (2016). How is the digital medium shaping research genres? Some cross-disciplinary trends. *ESP Today. Journal of English for Specific Purposes at Tertiary Level, 4*, 22–42.

Petroni, S. (2011). *Language in the multimodal web domain*. Aracne Editrice.

Puschmann, C. (2015). A digital mob in the Ivory Tower? Context collapse in scholarly communication online. In M. Bondi, S. Cacchiani, & D. Mazzi (Eds.), *Discourse in and through the media: Recontextualizing and reconceptualizing expert discourse* (pp. 22–45). Cambridge Scholars Publishing.

Sales, H. E. (2006). *Professional communication in engineering*. Palgrave Macmillan.

Sancho Guinda, C. (2022, forthcoming). From isolation to constellation: Narrative focalisation in technology disclosures. *Ibérica, 44*.

Stocklmayer, S. (2013). Engagement with science: Models of science communication. In J. K. Gilbert & S. Stocklmayer (Eds.), *Communication and engagement with science and technology: Issues and dilemmas. A reader in science communication* (pp. 19–38). Routledge.

Swales, J. M. (1990). *Genre analysis: English in academic and research settings*. Cambridge University Press.

Swales, J. M. (2004). *Research genres. Explorations and applications*. Cambridge University Press.

Swales, J. M., & Feak, C. B. (1994). *Academic writing for graduate students: Essential tasks and skills*. The University of Michigan Press.

Tardy, C. M. (2016). *Beyond convention: Genre innovation in academic writing*. University of Michigan Press. https://doi.org/10.3998/mpub.5173647

Wenger, E. (1998). *Communities of practice. Learning, meaning, and identity*. Cambridge University Press.

Zhang, Y. (2019). Scifotainment: Evolving multimodal engagement in online science news. In C. Sancho Guinda (Ed.), *Engagement in professional genres* (pp. 244–258). John Benjamins.

Zhang, Y., & O'Halloran, K. L. (2014). From popularization to marketization: The hypermodal nucleus in institutional science news. In E. Djonov & S. Zhao (Eds.), *Critical multimodal studies of popular discourse* (pp. 160–177). Routledge.

4

Dissemination of Knowledge During the COVID-19 Pandemic: A Conceptual Metaphor Analysis of Research and Popular Articles

Fatma Benelhadj

1 Introduction

Science is a "social activity" largely based on "good communication", as it can evolve only through a process of "criticism and testing" (Hermann Bondi, as cited in Day and Gastel, 2016, p. xv). A good communication of science is important both for the recognition of the researcher(s) and the authentication of original research. In fact, "the cornerstone of the philosophy of science is based on the fundamental assumption that original research *must* be published; only thus can new scientific knowledge be authenticated and then added to the existing database that we call scientific knowledge" (Day and Gastel, 2016, p. xv).

To be worthy of publication, a scientific paper needs to carry 'new' knowledge. When new ideas are being stated, "clarity is essential" (Day &

F. Benelhadj (✉)
LAD, Faculty of Letters and Humanities, University of Sfax, Sfax, Tunisia
e-mail: fatma.benelhadj@flshs.usf.tn

R. Plo-Alastrué, I. Corona (eds.), *Digital Scientific Communication*,
https://doi.org/10.1007/978-3-031-38207-9_4

Gastel, 2016, p. 3). The language should be "straightforward and unambiguous" to guarantee a full understanding of the scientific content by the readers (Karska & Prażmo, 2017, p. 102). However, explaining new ideas can be sometimes challenging to the writer, as new concepts are usually "poorly (or partially) understood [...] complex and often inaccessible to the senses" (Semino, 2008, pp. 131–132). One of the ways to simplify these new domains to the reader is through comparing them to other better-known domains through the use of metaphor.

2 Conceptual Metaphor

In their seminal work *Metaphors We Live By*, Lakoff and Johnson (1980) defended the claim that metaphor is conceptual, that is, occurring at the level of thought and pervasive in everyday language. They (metaphors) structure "how we perceive, how we think, and what we do" (Lakoff & Johnson, 1980, p. 4) through a set of mappings from a source domain (SD) on a target domain (TD). The TD represents the domain that the speaker wants to talk about, while the SD is more concrete and known within the community of practice. SDs, like the human body, animals, plants, buildings, machines, money, and journeys, among others, are used to structure more abstract domains like time, emotions, human relationships, life, death, morality, and thought. For example, the conceptual metaphor LIFE IS A JOURNEY has several mappings like: THE PERSON LEADING A LIFE IS A TRAVELER, LIFE STATES ARE LOCATIONS, AND LONG-TERM PURPOSES ARE JOURNEY DESTINATIONS. These mappings are realised in language via metaphorical expressions like: "He just *sails through* life" and "He saw teaching as just a *stopover* on his *way* to bigger things" (Lakoff et al., 1991, pp. 36–37).

SDs help speakers think and act towards the TD. In fact, Lakoff and Johnson (1980) posit the hypothesis that if the concept of ARGUMENT was structured by the SD DANCE instead of WAR, then all the experiences with arguments would have been different (pp. 5–6). The choice of the SD is, therefore, important from a discourse analytic point of view, as it reveals the speaker's way of framing the TD. Semino (2008) adds that "metaphors are seldom neutral" and they "often [include] specific attitudes and evaluations" (p. 32). This results from the hiding/highlighting principle of conceptual metaphors (CMs). For instance, in the ARGUMENT IS WAR metaphor, "the competitive, aggressive and confrontational aspects of arguments" are highlighted, while "the cooperative, constructive aspects" are hidden (Semino, 2008, p. 33).

In addition to the major function of being able to talk and think about abstract domains, metaphors can have other more specific functions. Among these, Goatly (1997) mentioned "filling lexical gaps", "cultivating intimacy", and "textual structuring" (pp. 148–167), which he classified following Halliday and Matthiessen's (2014) three metafunctions (ideational, interpersonal, and textual). Such functions are related to the context where they occur, and this led later studies to consider metaphors in specific discourses. A conceptual metaphor in a literary text would function differently in a political text, for instance, because of the different goals and purposes of each discourse. The classifications of metaphors by metafunctions help link the choices in the text to the context (register and genre).

In the present chapter, the focus will be on scientific discourse by comparing metaphors in two key genres: the Research Article (RA) and the Popular Article (PA). The RA is the genre used to introduce "new findings and claims in the community" (Koutsantoni, 2006, p. 20). The PA rewrites these original findings in a journalistic way. Although both genres present almost the same findings, their readers, structures, and styles are different (Swales, 2004, p. 217). Science news, studied in this chapter, is one of the PA genres available online. It re-contextualises the major findings of RAs, by making the scientific findings 'newsworthy' (Allan, 2009, p. 6).

3 Scientific Discourse

The language of science is "no unitary or stable thing" (Bazerman, 1988, p. 16). There are significant differences between disciplines and genres. Each discipline is "a tribe ... with its particular norms, nomenclature, bodies of knowledge, sets of conventions and modes of inquiry" (Hyland, 2004, p. 8). The medical science studied in this chapter focuses on the human body and on maintaining/restoring a healthy state. In medical science, case studies are observed and experimentations can be made, thus making it a positivist/scientific research paradigm (Gotti & Salager-Meyer, 2006, p. 10).

Through the research journey, medical scientists come across different genres where writing about their research is required. Research articles (RA) are produced at "advanced stages of individuals' enculturation in disciplinary communities" and they "present original research" (Koutsantoni, 2006, p. 23). RAs are published after going through a long process of editing and proofreading for both form and content. The form is very specific as it is defined by "tradition, editorial practice, scientific ethics, and the interplay of printing and publishing procedures" (Day & Gastel, 2016, p. 16). Experimental RAs usually follow the IMRD structure.

This genre is, however, not stable. One of the factors that has influenced academic publishing is digitality. First, the emergence of Web 1.0, characterised by "information distribution and retrieval" (Vettorel, 2014, p. 486), led to the creation of 'e-journals', which made science accessible to everyone in the world (Swales, 2004, p. 217). Second, Web 2.0 has emerged adding "(active) participation in social networks, media-sharing websites, wikis, and blogs, by user-generated content and sharing practices that are often integrated and hybridised in these digital environments" (Vettorel, 2014, p. 486). These practices influenced the dissemination of science, and journalists tend to re-contextualise the latest findings via social networks, in the form of popular articles (PAs) to be read by the non-specialists.

PAs can be classified as a hybrid genre, with hybridity being "an umbrella term for all kinds of blending, mixing, and combining that occur in genres and texts" (Mäntynen & Shore, 2014, p. 738). This hybrid genre mixes features of the RAs with those of "films, television documentaries and dramas, newspaper articles, reports and feature articles in magazines and so on" (Yiqiong, 2012, p. 1). The scientific content of the RA is taken as a basis for the production of a new type of text. In science news, more particularly, what is important is not the 'relevance' of the study to previous literature or the 'novelty' of the methods used (Weingart, 2012, p. 18) but rather the newness of the event reported. The text should present "a good science story" (Allan, 2009, p. 6) that can be shared later with friends on social media (Hermida, 2010, p. 88).

This transposition of texts from a scientific background to a popular readership entails a change in the use of language. Giannoni (2008) studied the popularising features of two unrelated domains (medicine and applied linguistics) and concluded that seven features are involved in the process of popularising English journal editorials: questions, metaphors, marked lexis, humour, personalisation, appeals to reader, and contingency (p. 216). Among these features, metaphors seem to play an important role, in addition to questions, in the process of popularising journal editorials (Giannoni, 2008, p. 216). These metaphors seem to originate from different registers like general English, colloquial English, journalistic or literary registers (Giannoni, 2008, pp. 220–221).

4 Metaphor in Science

PAs seem to inherit metaphors from the original Ras, on the one hand, and from general English and the media, on the other. In RAs, where science is diffused to specialists, metaphors play a crucial role in the "teaching or explication of theories" (pedagogical/exegetical metaphors) and in "expressing theoretical claims" (theory-constitutive) (Boyd, 1993, pp. 485–486). Boyd (1993) gives the example of "the description of the spatial localization of bound electrons in terms of an 'electron cloud,' or the description of atoms as 'miniature solar systems'" (p. 486) to illustrate the pedagogical metaphors, while an example of theory-constitutive metaphors is talking about "cognitive psychology" through the domain of "computer science and information theory," where the brain is conceived as a computer (ibid.).

These two functions may overlap in the sense that the same metaphor can be used "for theory-constructive and for pedagogical purposes" (Knudsen, 2003), in addition to the possibility of co-existing "with other functions, such as argumentation, persuasion, vividness, humour and so on" (Semino, 2008, pp. 133–134). In scientific discourse, metaphors are subject to the hiding/highlighting duality, which would influence the questions, the hypotheses, the explanations and the conclusions in that research (ibid.). In this sense, the dominant metaphorical source domains might structure the studied phenomena. In medical science, common SDs might be: BALANCE, WAR/PHYSICAL CONFLICT, COMMUNICATION, HUMAN BEINGS, AND MACHINES, according to a study by Semino (2008, p. 159).

SDs in expert genres are used in a conventionalised way to the point that they become 'closed' (Knudsen, 2003), that is, overused by the specialists to the point it is classified as a non-metaphorical scientific concept. These metaphors can be a basis for non-expert genres. In the process of popularising those scientific findings, metaphors lose their function of constructing the theory as "the purpose of the communicative interaction is not to develop this specialised knowledge, but to make it available to non-experts" (Engberg, 2020, p. 175). With the new context, metaphors gain a new pedagogical function, and therefore, "re-open" and are "explicitly represented in the text as *metaphors* [emphasis added]" (Knudsen, 2003, p. 1255), by being further explained to the lay reader. The source domains are more extensively and creatively used in the popular genres (Semino, 2008, p. 144).

Despite the fact that PAs inherit certain metaphors from expert writings, they are also influenced by the metaphors used in the media. In a study of the role of metaphor in the dissemination of knowledge in newspaper articles, Williams Camus (2015) emphasises that the relation between science, the media, and the public is not unidirectional because science and journalism are totally distinct, and therefore, the way science is presented would be different across these registers. This view is also held by Semino (2008) who claims that the use of metaphor in popular genres is similar to that in the media rather than to expert genres. PAs are influenced by both: the scientific content and the journalistic features.

One of the issues that has received much attention from science and the media is the COVID-19 pandemic. It is one of the TDs that was conceptualised in terms of different SDs in the media. Pérez-Sobrino et al. (2022) talk about "a *surge* of creative metaphoric expressions in a variety of communicative settings [emphasis added]" used to describe the pandemic (p. 127). This study revealed that during the first six months of the pandemic, different metaphors were used (WAR, FIRE, ENEMY, etc.) and that "each metaphor highlights some aspects of the relevant topic or experience and backgrounds others" (p. 127). A #ReframeCovid campaign was launched as a result of the use of the WAR metaphor, which has "been widely criticised [...] for inappropriately personifying the virus as a malevolent opponent, creating excessive anxiety, potentially legitimising authoritarian governmental measures, and implying that those who die did not fight hard enough" (Semino, 2021, p. 50). Other metaphors (JOURNEY, MONSTER, STORM, TSUNAMI, FIRE, etc.) (Pérez-Sobrino et al., 2022; Wicke & Bolognesi, 2020) were used in the media to overcome the negative implications of the WAR metaphor.

The present chapter seeks to study the metaphors used in the dissemination of medical knowledge about the COVID-19 pandemic and their functions in the genres of RAs and PAs. It also seeks to identify the influence of the expert writings and of the media on this hybrid genre of science news as far as the use of metaphor is concerned.

5 Methods

The corpus is a collection of RAs and PAs extracted from *sciencedaily* (https://www.sciencedaily.com/), an online magazine representing the latest research news for the non-specialists. It is also known for publishing articles taken from news releases by leading universities. The website provides a link to the original research article, on which the news release is based. Table 4.1 summarises their distribution.

The corpus is a collection of 16 popular articles selected and downloaded from the website on December 6, 2021. The selection was based on two criteria: that it deals with COVID-19 (thus published after the outbreak of the virus) and that it has a medical content: based on a published research article in the field of medicine. The corresponding research articles (also 16) are also collected (Table 4.1).

Table 4.1 Corpus distribution

	PA	*RA*	*Total*
Number of items	16	16	32
Number of words	12,667	69,107	81,774

After collecting the corpus, the identification of the metaphorical expressions was based on the Metaphor Identification Procedure (MIP) developed by the Pragglejaz group (2007). First, the texts are read for a better understanding of the general meaning. Second, the lexical units are determined. Third, the incongruent units are highlighted, and finally they are marked as metaphorical if they have other more basic meanings in other contexts. The metaphorical words are then arranged by theme and then for each theme a conceptual metaphor of the form X IS Y is proposed.

The metaphorical expressions used in the analysis are assigned codes. PA or RA stand for Popular Article or Research Article, respectively. Within the RA, other classifications are made: INT for introduction, MET for methods, RES for results, DIS for discussion and conclusion, title, and abstract.

Seeing the large number of metaphorical expressions in the corpus, the analysis will focus on the most frequent ones. In order to study the functions played by metaphors in the RAs and PAs, this study relies on Halliday and Matthiessen's (2014) three metafunctions: textual, ideational, and interpersonal. In fact, Goatly (1997), who accounted for the different functions of metaphors in discourse, claimed that Halliday and Matthiessen's classification (2014) links genre to the type of metaphor.

6 Results and Discussion

The analysis has revealed that the most frequent metaphors are almost similar: PERSONIFICATIONS, CONTAINER METAPHORS, AND EVENT STRUCTURE. In order to study the functions of these metaphors in their respective genres, the textual, ideational, and interpersonal metafunctions of these metaphors are explored.

6.1 Textual Function

Metaphors performing the textual metafunction in the corpus aim to build cohesion and coherence in the text in relation to the context

(Halliday & Matthiessen, 2014). Cohesion refers to the "set of lexicogrammatical systems that have evolved specifically as a resource for making it possible to transcend the boundaries of the clause" (Halliday & Matthiessen, 2014, p. 603), that is, combining clauses together to form a coherent text. Two main metaphors perform this function in the corpus: CONTAINER metaphors and PERSONIFICATION.

CONTAINER metaphors, found in both genres, can help the text be more cohesive, as in:

(1) ***PA 7****. In the published studies*, health care providers collected nasal and throat samples—from patients and noted the time of virus exposure or symptom onset and sample collection.
(2) ***RA 9. INT****. In this study*, we present an integrative network medicine platform that quantifies the association of COVID-19 with other diseases across six categories, including autoimmune, malignant, cardiovascular, metabolic, neurological, and pulmonary (Fig. 1C).
(3) ***RA 7. MET****.* The material cross-sections were derived from a number of sources and was based on dry air, these are summarised *in Table 1*.
(4) ***RA 7. RES****.* The spatially varying intensity of the far-UVC feld produced by the lamp is presented *in Fig. 1*.

The reference in example (1) from a PA shows that the whole sentence (presenting the research methods) is contained within '*the published studies*'. What is important in popular writing is to prove that the story presented originates from scientific material. In the RA (example (2)), containment is made within the same study '*in this study*' and can be linked to its position at the end of the introduction section. In fact, after reviewing the literature necessary for their research, the author moves to state the focus of their research—"to occupy the niche" (Swales, 2004, p. 231). In this way, the use of the containment guides the reviewer through the different sections of the article.

The CONTAINER metaphor is further explored in the RA. Other, more specific, containers are mentioned in examples (3) and (4): *Table 1 and Fig. 1*. Tables and figures are important in presenting data in medical research, and are not reproduced in PAs. The textual references help establish the link between the table or the graph and the author's comment or explanation. Containment, in this case, helps the author express their own voice and be more precise and specific.

The second metaphor that plays a textual function is PERSONIFICATION. In the RAs, the study/research is conceptualised as a person performing actions like to '*show*' or '*suggest*':

(5) ***RA. INT. 11:*** Our study *shows* that the specific combination of tumor necrosis factor a (TNF-a) and interferon g (IFN-g) is critical for these processes.
(6) ***RA.DIS. 3:*** A recent study using injected microbubbles *suggests* that vasodilation occurs in some mechanically ventilated patients with COVID-19 and is correlated with PaO2:FiO2.

In example (5), conceptualising the '*study*' as a person allows the researcher not to directly refer to themself and to build the coherence of the text by referring to the whole publication. After presenting the literature, this metaphor helps announce the focus of the research and paves the way to discover the methodology adopted. The textual function is achieved differently in example (6). In fact, what is conceptualised as a person '*suggesting*' ideas is actually another study. The textual reference is made to another text and not to a section of the same text. The occurrence of this metaphor in the discussion section of an RA is very important as it helps the researcher discuss their findings with respect to other researchers'.

These examples are compared to other instances from PAs:

(7) ***PA 1. TITLE.*** COVID-19 virus enters the brain, research strongly *suggests*—A new study *shows* how spike protein crosses the blood-brain barrier.
(8) ***PA 13.*** The study also *demonstrates* the power of frequent testing in shortening the pandemic and saving lives.

Similar to RA, the '*study*' or '*research*' is conceptualised as a PERSON. What is different, however, is that this person is also able to '*demonstrate*', which means that the study is not conceptualised as a PERSON in general, it is more specifically a RESEARCHER. Added to that, the textual function that these metaphors play is to link the PA to another text: the original RA.

Structurally speaking, using the reporting verb at the end of the clause (example 7: 'research strongly *suggests*') is frequent in news reporting rather than academic writing (Biber et al., 2007, p. 925). Placing the reporting verb at the end of the clause is a structural variation where "the quoted text is the main communicative point, and the reporting clause is tagged on at the end" (Biber et al., 2007, p. 924). Using this strategy in most of the titles (6 out of 16) shows the extent to which PA authors want to reassure their readers about the credibility of their story, and therefore, attract them to read the text. PERSONALISATION in this way transcends the textual metafunction to perform the interpersonal metafunction of persuasion.

CONTAINER metaphors and PERSONIFICATION play an important role in linking the text together. Still this function varies from one genre to the other, for while in RA, these metaphors make references more precise and help the authors build their persona, in PA, they highlight the credibility of the news.

6.2 Ideational Function

The ideational metafunction is about how language is used to represent the world (Halliday & Matthiessen, 2014). More specifically, it represents "what/who did what to what/whom" by identifying the '**participants**' who are performing a '**process**' in a particular '**circumstance**' (Thompson, 2014, p. 91). Metaphor gives the possibility to abstract concepts to be treated in the clause as a participant through the process they are associated with. One of the metaphors performing this function is PERSONIFICATION, as in the following example:

(9) ***RA 1.TITLE.*** The S1 protein of SARS-CoV-2 *crosses* the blood–brain barrier in mice

It is unclear whether severe acute respiratory syndrome coronavirus 2, which causes coronavirus disease 2019, *can enter* the brain. Severe acute respiratory syndrome coronavirus 2 *binds* to cells via the S1 subunit of its spike protein. We show that intravenously injected radioiodinated S1 (I-S1) readily *crossed* the blood–brain barrier in male mice, *was taken* up by brain regions and *entered* the parenchymal brain space. I-S1 was also *taken* up by the lung, spleen, kidney and liver. Intranasally administered I-S1 also *entered* the brain, although at levels roughly ten times lower than after intravenous administration. ..... Mechanistic studies indicated that I-S1 *crosses* the blood–brain barrier by adsorptive transcytosis and that murine angiotensin-converting enzyme 2 is involved in brain and lung uptake, but not in kidney, liver or spleen uptake.....Therefore, we next assessed whether intravenously injected I-S1 fully *crosses* the capillary wall to enter brain parenchyma. ..... We found that binding of I-S1 to the luminal side *remained static* over time, whereas there was a small increase over time in the amount of I-S1 *retained* by brain capillaries (Extended Data Fig. 3). The largest change over time was in the amount of I-S1 *entering* the parenchyma. These results show that by 30 min, over 50% of I-S1 *had crossed* the capillary wall fully to *enter* into the brain parenchymal and interstitial fluid spaces.

This excerpt from a research article conceptualises 'the S1 protein of SARS-CoV-2' as a PERSON, and this metaphor is used all over the text: from the title to the conclusion. This can be detected from the processes associated with it: '*cross, enter, bind, be taken and be retained*' (italics in the example). However, focusing on the circumstances linked to these verbs, it is possible to specify that the protein rather behaves like a burglar as it crosses 'the capillary wall' and it takes '30 min' to do so (underlined in the example). The metaphor SPIKE PROTEIN IS A BURGLAR runs throughout the whole text and helps explain the whole idea: the virus 'binds' to cells using this protein and can therefore 'enter' the body and 'cross the capillary wall' while trying to hide. The metaphor is both 'theory-constitutive' as it helps develop the theory built by the researcher, and 'pedagogical' as it explains the general idea of the text.

In the PA, the spike protein is conceptualised as a MONSTER:

(10) ***PA 1***. In a study published Dec.16 in Nature Neuroscience, researchers found that the spike protein, often *depicted* as the *red arms* of the virus, can *cross* the blood-brain *barrier* in mice. This strongly suggests that SARS-CoV-2, the cause of COVID-19, *can*

enter the brain. The spike protein, often called the S1 protein, *dictates* which cells the virus *can enter*. Usually, the virus *does* the same thing as its binding protein, said corresponding author William A. Banks, a professor of medicine at the University of Washington School of Medicine and a Puget Sound Veterans Affairs Healthcare System physician and researcher. Banks said binding proteins like S1 usually by themselves *cause damage* as they *detach* from the virus and *cause* inflammation.

The author of this PA uses almost the same metaphorical expressions as the RA, namely, '*can cross the blood-brain barrier*' and '*can enter the brain*', among others. However, the whole concept seems to be different. In fact, the image of the virus binding to cells via the protein in order to cross the wall is lost, and instead, the virus is conceptualised as a MONSTER having '*red arms*', '*dictating*' which cells to enter, and '*causing damage*'. This metaphor does not fully explain the mechanism of entering the body as it has lost some of its 'theory-constitutive' function. The explanatory function, however, is fulfilled, but with a potential to frighten the audience, as such creatures (or monsters) have a 'destructive potential' and can convey "a sense of threat" (Cassese, 2018, p. 9). In this case, the metaphor can be said to perform a double function: ideationally, it represents what is going on, and interpersonally it is trying to influence the reader. This metaphor is largely used in the media (Cassese, 2018). This is further stressed by other metaphors:

(11) ***PA 3.*** In coronavirus patients, it's thought that the infection first *damages* the lungs, *rendering* parts of them incapable of functioning properly.
(12) ***RA 3.*** As critical illness in patients with COVID-19 is caused, in part, by inflammatory injuries that *affect* the lungs and lung blood vessels…
(13) ***PA 6.*** COVID-19 is *making* tinnitus *worse*
(14) ***RA 14.*** The pandemic has been shown to increase fear and worry in the general population (14) and may potentially *worsen* levels of tinnitus due to the clear relationship between emotional distress and severe tinnitus (15).

(15) ***PA.14. Title.*** COVID-19 infection combined with blood clots *worsens* patient outcomes, study finds

PA 3 uses the term '*damage*', which is a hyponym of '*affect*', used in RA 3. More meanings and richer images are provided in the popular writing, first, to better explain the notions dealt with and, second, to influence the behaviour of the reader. Similarly, the term '*worse*' is hedged in RA 14 *(may potentially)*, while it is stated blatantly in the title of the PA (example 15), thus highlighting the agentivity of the virus. These metaphors depict COVID-19 as an ENEMY that needs to be killed:

(16) ***PA 11.*** COVID-19 is caused by the SARS-CoV-2 virus. The infection has *killed* more than 1.2 million people in less than one year and sickened millions more.
(17) ***PA 1.*** The immune system, upon seeing the virus and its proteins, overreacts in its attempt to *kill* the *invading* virus.
(18) ***PA 16.*** Research is paving the way for a new solution to *kill* aerosolized COVID-19 in enclosed environments such as hospitals and long-term care facilities.
(19) ***RA 16. ABSTRACT.*** To *reduce* aerosol *transmissions*, the use of low dose far-UVC lighting to disinfect in-room air has been proposed.

COVID-19 is an INVADER who killed, because of the infection it causes, millions of people. This justifies the action of researchers who aim to '*kill*' the '*aerosolized COVID-19*'. The image is different in RA 16, as the author deals with '*reducing the transmission of aerosol COVID-19*'. COVID-19 is described as a commodity that can be 'transmitted'.

In addition to the virus itself, the patients are one of the subject matters in scientific publication during the pandemic. They are conceptualised as CONTAINERS:

(20) ***RA 3.*** … the large amount of pulmonary venous admixture observed in patients with early COVID-19.. Early reports of the disease caused by the novel coronavirus (COVID-19) describe at least *a subset of patients* who present with hypoxemia
(21) ***RA 4.*** As critical illness *in patients with COVID-19* is caused, in part, by inflammatory injuries

(22) ***RA 11.*** Inflammatory Cytokines Are Elevated *in Patients with COVID-19*
(23) ***PA 7.*** Kucirka says patients who *have* a high-risk exposure should be treated as if they are infected...
(24) ***PA 8.*** Almost one in five patients with COVID-19 *may only show* gastrointestinal symptoms...
(25) ***PA 11.*** St. Jude Children's Research Hospital scientists have discovered a potential strategy to prevent life-threatening inflammation, lung damage and organ failure *in patients with COVID-19.*

In RAs, the term 'patient' collocates most of the time with the preposition 'in', thus creating a CONTAINER metaphor: the patients are containers for different medical phenomena. In PAs, it is possible to depict the patient as 'having a high-risk exposure', 'showing symptoms' and also CONTAINERS (example 25). This difference can be linked to the conceptualisation of patients in medicine, who are depicted as "the 'work object' of medicine" (Armstrong, 2002, p. 245). Traces of this metaphor can be found in PAs (example 26) but are not frequent. Patients are rather conceptualised as AGENTS capable of doing actions. In popular writing, this metaphor emphasises the humanity of the patients and highlights their struggle against the disease, thus convincing the readers that they can be AGENTS and not only patients of COVID-19.

Although the PAs inherited their subject matter from the RA, they still have their own additions to the story being told first in terms of explanations (more specific meanings) and second in terms of conceptualisation of participants.

6.3 Interpersonal Function

The interpersonal function concerns how to 'interact' with people using language or how "to establish and maintain appropriate personal and social link with them" (Thompson, 2014, p. 45). Metaphors in science writing can help establish a relation between the author (researcher or journalist) and reader (specialist or layperson). References to the authors are analysed in these examples:

(26) ***RA 1. ABSTRACT.*** We *show* that intravenously injected radioiodinated S1 (I-S1) readily crossed the blood–brain barrier in male mice
(27) ***RA 3. MET.*** In the present study we *use* a mathematical model of perfusion and oxygen transport to address these questions.

In RAs, researchers are able 'to show' things that normal people do not come to perceive (example 26) and 'use' mathematical models to prove what they say (example 27). These actions are classified as metaphorical as a result of conceptualising phenomena as OBJECTS and therefore use them as part of the EVENT STRUCTURE METAPHOR. The researchers are conceptualised as TRAVELLERS through the research process. Based on the actions performed, the researcher seems to abide by the norms of academic writing. The actions of 'showing' and 'using' are related to the sections where they are mentioned: the abstract, where the researcher needs to argue for the validity of their claim, and the methods, where they need to identify the tools used in research.

In the PA, reference to first person pronoun is generally made when the researcher is given a space to present their research.

(28) ***PA 7.*** "We are *using* these tests *to rule out* COVID-19, and basing decisions about what steps we *take* to prevent onward transmission, such as selection of personal protective equipment for health care workers," says Kucirka. "As we *develop* strategies to reopen services, businesses and other venues that rely on testing and contact tracing, it is important to understand the limitations of these tests."
(29) ***PA 14.*** "We'*re still in* the process of understanding the pathophysiology of COVID-19, so it's important to have a low index of suspicion when it comes to this infection to ensure we'*re doing* all we can to *mitigate* the spread and prevent severe outcomes."
(30) ***PA 16.*** "We found that far-UVC illumination in poorly ventilated spaces can be as effective as N95 masks in preventing transmission. With detailed and accurate computational fluid dynamics modelling, we *were able to track* and *eliminate* the airborne transmission of pathogens."

These instances are quotes from interviews with researchers. The researchers in these cases seem to perform other actions: '*using* tests to *rule out* COVID-19', 'what steps we *take* to prevent', 'be able to track', and so on. The researchers in these examples perform the role of the SAVIOURS of humanity. In fact, the image of being 'a humble servant of the discipline', perceived in the RA, is complemented by the need to "affirm a commitment to his or her work" (Hyland, 2001, p. 224).

7 Conclusion

This chapter has tried to study how medical findings about COVID-19 have been disseminated during the pandemic. Popular articles, produced during the pandemic, and their corresponding research articles have been studied from a conceptual metaphor perspective. The analysis revealed that while the general use of metaphors can be similar across the genres, their functions are rather different and genre-driven. In RAs, metaphors can add to the cohesion of the text: they might be used to make more specific textual references through containment and to avoid personal references through personification. They are also thought to play a double role in the studied corpus: both theory-constitutive and pedagogical functions are performed by the same CMs. Additionally, metaphors can reveal the identity of the authors as 'servants of the discipline' and the patients as subject of study. In PAs, however, which rely on RA for their stories, metaphors help build the cohesion of the text but perform a second function of persuading the readers of the credibility of their stories. Also, while relying on the metaphors used in their respective RAs, PA authors opt for more explanations through adding more vivid images in the metaphors and neglecting certain details. Moreover, unlike RAs, PAs highlight the humanity of the patients and give more space to the researchers to identify themselves as saviours.

Following this, the relation between the two genres can be said to be complementary rather than unidirectional, for while PAs rely on RAs for their stories, they add to these stories and they offer to scientists the space to express their identities. PAs are not just reproductions of RAs, they actually use the scientific discoveries to insinuate the metaphors used in the media (the MONSTER metaphor). Furthermore, with the social media and the sharing practices, the human facet of the researchers is gaining space.

References

Allan, S. (2009). *Making science newsworthy: Exploring the conventions of science journalism* (pp. 149–165). Oxford University Press.
Armstrong, D. (2002). Conceptualising the patient. *Journal of Health Services Research & Policy, 7*(4), 245–247.
Bazerman, C. (1988). *Shaping written knowledge: The genre and activity of the experimental article in science*. University of Wisconsin Press.
Biber, D., Johansson, S., Leech, G., Conrad, S., Finegan, E., & Quirk, R. (2007). *Longman grammar of spoken and written English* (Vol. 2).
Boyd, R. (1993). Metaphor and theory change: What is 'metaphor' a metaphor for? In A. Ortony (Ed.), *Metaphor and thought* (2nd ed., pp. 481–533). Cambridge University Press.
Campagna, S., Garzone, G. E., Ilie, C., & Rowley-Jolivet, E. (2012). *Evolving genres in web-mediated communication*. Peter Lang.
Cassese, E. C. (2018). Monster metaphors in media coverage of the 2016 US presidential contest. *Politics, Groups, and Identities, 6*(4), 825–837. Taylor & Francis.
Day, R. A., & Gastel, B. (2016). *How to write and publish a scientific paper*. Cambridge University Press.
Engberg, J. (2020). Institutional dissemination of legal knowledge: An instance of knowledge communication. In M. Gotti, S. Maci, & M. Sala (Eds.), *Scholarly pathways: Knowledge transfer and knowledge exchange in academia* (pp. 175–205). Peter Lang.
Giannoni, D. S. (2008). Popularizing features in English journal editorials. *English for Specific Purposes, 27*(2), 212–232.
Goatly, A. (1997). *The language of metaphors*. Routledge.
Gotti, M., & Salager-Meyer, F. (2006). *Advances in medical discourse analysis: Written and oral contexts*. Peter Lang.
Halliday, M. A. K., & Matthiessen, C. (2014). *An introduction to functional grammar*. Routledge.

Hermida, A. (2010). Revitalizing science journalism for a digital age. In D. Kennedy & G. Overholser (Eds.), *Science and the media* (pp. 80–87). The Media in Society Project.
Hyland, K. (2001). Humble servants of the discipline? Self-mention in research articles. *English for Specific Purposes, 20*(3), 207–226.
Hyland, K. (2004). *Disciplinary discourses, Michigan classics ed.: Social interactions in academic writing*. University of Michigan Press.
Karska, K., & Pražmo, E. (2017). Didactic potential of metaphors used in medical discourse. *Linguistics Beyond and Within (LingBaW), 3*, 102–116.
Knudsen, S. (2003). Scientific metaphors going public. *Journal of Pragmatics, 35*(8), 1247–1263.
Koutsantoni, D. (2006). Rhetorical strategies in engineering research articles and research theses: Advanced academic literacy and relations of power. *Journal of English for Academic Purposes, 5*(1), 19–36.
Lakoff, G., Espenson, J., & Schwartz, A. (1991). Master metaphor list: 2nd draft copy. *Henta fra:* http://araw.mede.uic.edu/alansz/metaphor/METAPHORLIST.pdf, *16*, 80.
Lakoff, G., & Johnson, M. (1980). *Metaphors we live by*. University of Chicago Press.
Mäntynen, A., & Shore, S. (2014). What is meant by hybridity? An investigation of hybridity and related terms in genre studies. *Text & Talk, 34*(6), 737–758.
Pérez-Sobrino, P., Semino, E., Ibarretxe-Antuñano, I., Koller, V., & Olza, I. (2022). Acting like a hedgehog in times of pandemic: Metaphorical creativity in the# ReframeCovid collection. *Metaphor and Symbol, 37*(2), 127–139.
Pragglejaz Group. (2007). MIP: A method for identifying metaphorically used words in discourse. *Metaphor and Symbol, 22*(1), 1–39. https://doi.org/10.1080/10926480709336752
Semino, E. (2008). *Metaphor in discourse*. Cambridge University Press.
Semino, E. (2021). 'Not soldiers but fire-fighters'–Metaphors and Covid-19. *Health Communication, 36*(1), 50–58.
Swales, J. M. (2004). *Research genres: Explorations and applications*. Cambridge University Press.
Thompson, G. (2014). *Introducing functional grammar*. Routledge.
Vettorel, P. (2014). Evolving Genres in Web-mediated Communication. *ELT Journal, 68*(4), 486–490.

Weingart, P. (2012). The lure of the mass media and its repercussions on science. In S. Rödder, M. Franzen, & P. Weingart (Eds.), *The sciences' media connection–public communication and its repercussions* (pp. 17–32). Springer.
Wicke, P., & Bolognesi, M. M. (2020). Framing COVID-19: How we conceptualize and discuss the pandemic on Twitter. *PLoS ONE, 15*(9), e0240010. https://doi.org/10.1371/journal.pone.0240010
Williams Camus, J. T. (2015). Metaphor, news discourse and knowledge. In *Metaphor in specialist discourse* (pp. 245–269). John Benjamins Publishing Company.
Yiqiong, Z. (2012). Representing science: A multimodal study of science popularization on institutional and mass media websites. (ScholarBank@NUS Repository. http://scholarbank.nus.edu.sg/handle/10635/36378).

5

Authorial Stance and Identity Building in Weblogs by Law Scholars and Scientists

Giuliana Diani and Maria Freddi

1 Introduction

The ways that writers and speakers express their opinions is an important feature of all interaction. As suggested by Hyland and Jiang (2019, p. 129), "successful research writers construct texts by taking a novel point of view towards the issues they discuss while anticipating readers' imagined reactions to those views. This intersubjective positioning is encompassed by the term *stance*." From this perspective, stance refers to "the ways writers present themselves and convey their judgements, opinions, and commitments" (Hyland, 2005a, p. 176). This notion has often been discussed in relation to the notion of identity as taking a stance can contribute to identity construction in the form of authorial self-mention,

G. Diani (✉)
Università degli Studi di Modena e Reggio Emilia, Modena, Italy
e-mail: giuliana.diani@unimore.it

M. Freddi
Università Vita-Salute San Raffaele, Milano, Italy
e-mail: freddi.maria@hsr.it

R. Plo-Alastrué, I. Corona (eds.), *Digital Scientific Communication*,
https://doi.org/10.1007/978-3-031-38207-9_5

writer visibility, and interaction with the reader. Our study adopts this perspective of stance and identity.

Studies of stance and identity in academic communication have highlighted the discursive deployment of epistemic and affective language conveying the author's perspective, their angle, but also their attitudes and sentiments (particularly in the wake of Hyland, 2000, 2005a, 2005b). In the argumentative genres at the core of written academic discourse, that is, the research article, but also the learner genres of Bachelor's, Master's, and Doctoral Theses, this authorial presence is marked by hedges, that is, ways of withholding complete commitment to a proposition by means of epistemic modals (Hyland, 2005b, p. 52), and complementary boosters, that is, ways of strengthening a claim using adverbs (stance adjuncts) of the *surely, no doubt* kind (Hyland, 2005b). The other obvious resources through which the author marks their presence in the text are first-person pronouns *I* and *we*, what Hyland (2001, 2005b) has called "self-mention", or, as recently rephrased in a study on digital genres, "the writers/speaker's intrusion in the text through use of first-person to emphasise their contribution" (Zou & Hyland, 2022, p. 227). A further category of stance markers includes the whole set of attitude markers falling on the affective end of the epistemic–attitudinal continuum and marking the speaker/writer's more personal beliefs and feelings of surprise, frustration, agreement, and importance, among others (e.g. first-person subject pronouns with verbs *hope*, *agree*, *prefer* as in *I hope/agree*, and adverbs such as *hopefully*, *surprisingly*, *unfortunately* [Hyland, 2005b]).

It is only recently that scholars have started looking at realisations of stance in the new digital genres that have become a vehicle for knowledge transfer to the wider community of the Internet, with both experts and non-expert readers. Luzón (2011, 2013), for example, explored academic blogs as a new form of scholarly communication which provides both a space for researchers to present their ideas, observations, and reactions to others' work and a discussion forum. With an interest in authorial stance in three-minute thesis presentations and academic blogs, Hyland and Zou (2021) and Zou and Hyland (2020, 2022) examined the ways doctoral students and academics present themselves and their research to non-specialist audiences in "public arenas" such as blogs. Considerable scholarly attention has also been paid to domain-specific blogs. Freddi (2020) examined interaction and audience engagement in science blogs written by individual scientist-writers. Garzone (2014, p. 167) investigated the personal individualistic dimension of law blogs or "blawgs", as "blogs used by the legal and professional community to exchange scholarly opinions, to debate topical issues and discuss important legal cases". Similarly, Tessuto (2015, p. 85) examined legal blogs as sites "for stance and engagement". Diani (2021) highlighted the argumentative dimension of law blogs.

In this chapter, we analyse legal and science blogging as knowledge exchange platforms for academics, students, and researchers. We focus on self-mention, first-person pronouns *I*, *me*, and possessive *my*, together with the verbs that most frequently co-occur with these author's pronouns, to map types of stance along the epistemic–attitudinal axis. In doing so, we identify similarities and differences between law and science bloggers in the way they mark their presence and position themselves in relation to their arguments and audience. In the next section, we describe the blogs used for the study and the methodology adopted. In Sect. 3, we present the results emerging from the comparative analysis of the two types of blogs, and finally, in Sect. 4, we discuss the findings and draw some conclusions.

2 Materials and Methods

2.1 Materials

The study is based on the analysis of two comparable corpora of blog texts in English: law blogs and science blogs, respectively, selected over a six-year period, from 2017 to 2012 (law) and 2019 and 2014 (science) in the typical reverse chronological order of blog posts.

The law blog corpus consists of 96 posts (totalling 175,849 words) taken from two legal blog websites: one from the UK (*United Kingdom Constitutional Law Association*, UKCLA), and one from the US, SCOTUSblog (see Table 5.1). Only blog posts commenting on legal

Table 5.1 Composition of comparable corpora: law and science blogs

Discipline	Blog	No. of posts	Tot. no. of tokens
Law	SCOTUSblog	50	77,648
	UKCLA	46	98,201
	Total	96	175,849
Science	Genomics, Medicine and Pseudoscience (GEN)	120	94,913
	Mountain Beltway (MB)	199	89,091
	NeuroLogicaBlog (NEU)	40	45,518
	Total	359	229,522

cases written by single-authored law professors and lecturers affiliated to UK and US universities were chosen. The choice of single-authored posts was made in order to avoid potential differences with respect to the strength and focus of the positions expressed, and to create a more clearly identifiable category of blog. The two websites were selected for their relevance within the academic legal community. UKCLA "is the UK's national body of constitutional law scholars affiliated to the International Association of Constitutional Law". As stated in the 'About us' homepage, "its object is to encourage and promote the advancement of knowledge relating to United Kingdom constitutional law and the study of constitutions generally". Its blog, launched in 2010 (https://ukconstitutionallaw.org/blog/), is "a repository of expert comment and analysis on matters of constitutional law in the UK and further afield". SCOTUSblog (http://www.scotusblog.com/), founded in 2002 by attorneys Tom Goldstein and Amy Howe, reports on cases of the Supreme Court of the US. The blog posts are written by lawyers, law professors, and law students who analyse the cases before the US Supreme Court and post breaking news of court decisions.

The science blog corpus contains 359 posts (229,522 words) taken from three US science blogs maintained by individual academics of different seniority, who not only are engaged in scientific research in the physical sciences or medicine and health, but also engage in public writing and use the blog as a platform for free expression and interaction on the web (see Table 5.1). The first blog sampled is *Genomics, Medicine and Pseudoscience* (hereafter GEN, http://genome.fieldofscience.com) run by

Steven Salzberg, a professor at Johns Hopkins School of Medicine. He deals with a range of topics in genomics and medicine as well as pseudoscience and unscientific medical practices as he writes in the blog's subtitle "Here's where I can say what I really think about abuses and distortions of science, wherever I see them". *Mountain Beltway* (MB, https://blogs.agu.org/mountainbeltway/) is about geology and earth sciences. The author is Callan Bentley, Associate Professor of geology at Piedmont Virginia Community College in Charlottesville and Fellow of the Geological Society of America. A similar disclaimer specifies that "ideas and opinions expressed on this site are those of the authors and commenters alone. They do not necessarily represent the views of the American Geophysical Union." The third blog chosen is *NeuroLogicaBlog* (NEU, https://theness.com/neurologicablog/) with the subtitle "Your daily fix of Neuroscience, Skepticism and Critical Thinking". This is written by Steven Novella, a clinical neurologist at Yale University School of Medicine, who is also the host of a science podcast and co-founder of the New England Skeptical Society. Although the three blogs are independently authored blogs, two of them are hosted respectively by a professional society (*Mountain Beltway* is hosted by the American Geophysical Union, AGU) and by an independent network of science blogs (Field of Science, hosting Steven Salzberg's blog). All three blogs are featured among the science blogs worth following by several blog feeds and networks (e.g. Ata Scientific).

In order to make the two corpora comparable, the same sampling criteria were applied: all bloggers are scholars directly involved in research. In the case of the law blogs, they are also practising lawyers. Their blog can thus be considered an example of *research blogging* with the double aim of advancing scholarship and committing to public engagement in science (as discussed in Mauranen, 2013). However, it must be pointed out that there are no British scholars among the science blogs sampled.

The readers of both types of blogs are mainly experts, as is apparent from the contents of the posts and the quality of the comments. One of the science blogs, however, namely, *NeuroLogicaBlog*, seems to attract a wider audience, judging by the readers' nicknames and by the quantity of comments. Although both blogs have comment threads, the law blogs have fewer compared to the science blogs (cf. Diani, 2022). For the purpose of comparability, therefore, only the posts are included in the analysis. Moreover, the law blogs comment on court decisions and, therefore, in the same time interval there are fewer posts with a higher number of words each, while the science blogs instead have almost four times as many posts with fewer words each (*NeuroLogicaBlog* again is an exception with 40 posts only but longish ones). The law posts sampled from UKCLA are written by more than 30 different bloggers, 6 of whom write more than one post over the decade. On the contrary, the posts selected from SCOTUSblog are written by the same four bloggers. This is also the case for the science blogs where there are only three individuals that keep posting on many different topics, often responding to scientific news, commenting on a book or other new publication, or just raising discussion on a controversial scientific issue, thus the posts overall are almost four times as many as in the law blogs. This difference in the composition of the corpora might help interpret some of the findings in Sect. 3.1 relative to the distribution of first-person pronouns and of the possessive.

2.2 Methods

From a methodological point of view, we conducted a frequency-driven analysis of word distributions in the two corpora, using the AntConc (Anthony, 2022) suite of programmes for electronic text analysis, in particular Word Lists and Concordance. We then analysed verbs collocating with first-person subject and object pronouns and the possessive *my* ("self-mention" as in Hyland, 2005b) by looking at each single concordance manually when the number of instances was manageable in size and by taking automatically calculated collocates when the instances were too many to handle manually (using AntConc Collocates). A subsample of concordances was analysed manually in this second instance.

As a further step, we took the frequency distribution of collocates and grouped the verbs into semantic categories of verbs (following Hyland, 2002, 2005b and Halliday & Matthiessen, 2014). We then considered the functions of each category by analysing the expanded concordances of *I* + verbs to explore the construction of stance (types of stance along the epistemic and attitudinal axis) and identity, and the different communicative functions of law and science blogs.

To retrieve self-mention, the automated search was used, for example, to separate instances of capitalised *I* versus *i* by adjusting the Word List settings in AntConc (select "Treat case in sort"), then in the Concordance search (select both Words and Case and search for the capitalised form). However, the automated search returned some instances that needed to be discarded manually, for example, *World War I* in the science blog sample, or *Section I* in the law blog corpus. This applied to just a limited number of instances, so they did not alter the distributions. In the law blog corpus, some of the occurrences of *I* were discarded because they are part of reported speech.

3 Results

3.1 An Overview of Subjectivity Markers

A preliminary overview of the quantitative data in the two blog corpora highlights bloggers' propensity for subjectivity and self-expression. However, frequency differences do seem to exist. More specifically, first-person singular subject pronoun *I* ranks 12th (1876 occurrences) in the science corpus, whereas it ranks 117th (178 occurrences) in the law blog corpus. Such a significant difference in the ranking might be due to the different types of blogs compared. While the science blog posts in the corpus are numerous and written by the same three scientists with the main purpose of expressing their thoughts on things they have read and seen, or reacting to some important story, the law posts are fewer and written by different bloggers whose aim is to analyse cases and comment on court decisions. This might explain why the first-person singular subject pronoun is pervasive in science blogging, but less so in legal blogging, where *I* does not rank among the top 20 words and the top 50 words in the word list are mainly legal terms like *Court, case, law, justice, Supreme, Lord*, and *decision*. In contrast, in the science blogs word list, the top 50 words are all grammatical. This might point to a more personal and colloquial style of blogs revolving around the scientist-blogger and his audience versus a more impersonal style of law blogs where the focus is the ideational content of the court decision, the reasoning, and legal argumentation. This is also confirmed by the very low frequencies of the object pronoun *me* and the possessive *my* in the law blogs: the possessive *my* ranks 512th (42 occurrences) and the object pronoun *me* ranks 1185th (18 occurrences), while in the science blogs the possessive *my* ranks 57th (453 occurrences) and the object pronoun *me* ranks 97th (264 occurrences). As Table 5.2 shows, there is a similar distribution of the three forms in the law and science blogs, *I* being followed by *my* and then *me*, but a significant difference in the frequency with which the subject pronoun *I* occurs across the two corpora (0.10% in law versus 0.82% in science).

These quantitative results concerning self-mention reflect the finalities of the blog posts and call for further analysis of the collocational profiles in the two types of blogs.

3.2 Collocational Profile of the Subject and Object Pronouns *I* and *Me* and the Possessive *My* in the Law and Science Blogs

In this section, we examine the collocational profile of the subject and object pronouns *I* and *me* and the possessive *my* in the law and science blogs considered separately. The collocates are illustrated through concordance lines, expanded examples, and summarised in collocate tables, where relevant to the discussion.

Table 5.2 Frequency distributions of *I*, *me*, and *my* in the law and science blogs (descending frequency order)

Self-mention	Law blogs		Science blogs	
	raw freq.	%	raw freq.	%
I	178	0.10	1876	0.82
my	42	0.02	453	0.20
me	18	0.01	264	0.12

Collocational Profile of Subject and Object Pronouns *I* and *Me* and the Possessive *My* in Law Blogs

In the law blogs, verbal word-forms unsurprisingly score among the top 20 most frequent collocates of *I* immediately to the right of the pronoun (R1 position). Apart from the link-verbs *have* and *be* (*was*, *am*), the most substantial group of collocates accounts for verbs having to do with mental processes, such as *think*, *believe*, *doubt*, *suspect*, *found* (in the sense of *believe*), *understand*. The verb *think* is attested as the first most frequent collocate (13 occurrences), followed by *believe* (4 occ.), *doubt* (4 occ.), *suspect* (3 occ.), *found* (3 occ.). Examples from the corpus are:

(1) *I think* the courts would be defeated, and in the end the Supreme Court would exercise its power under the Practice Statement of 1966 to reverse its position and reinstate the doctrine of parliamentary supremacy. (UKCLA)

(2) In the space below, I analyze the merits portion of Mr. Clement's brief on behalf of the Arizona legislature, and point out why *I think* it fails to demonstrate that the IRC's creation and powers violate federal law. (UKCLA)

(3) Justice Kennedy was surprisingly quiet at oral argument, asking just two questions about whether the identity of the party bound by the customary-international-law norm should be part of the analysis at step one. The question is a good one, and *I believe* the answer is yes—the particular norm of customary international law must apply to a corporation before a corporation may be sued under the ATS for violating that norm. (SCOTUSblog)

According to Hyland (2000, p. 123), verbs like *think*, *believe* allow writers to "build a personal ethos through an impression of certainty, assurance and conviction in the views expressed, an image strengthened with the use of personal pronouns". Examples (1) to (3) illustrate this certainty booster (to use Hyland's terminology) very well. Through *I think*, *I believe*, the blogger shows an overt acceptance of personal responsibility for their argument and makes the conclusion stronger (*I think the courts would be defeated; I think it fails to demonstrate that...; The question is a good one, and I believe the answer is yes*). The second most frequent set of word-forms includes verbs of saying or "discourse" verbs (Hyland, 2005a) such as *discussed* (5 occurrences), *argue* (5 occ.), *offer* (5 occ.), *wrote* (5 occ.). As already shown by Diani (2021), the use of discourse verbs in the scholarly legal blog genre may be related to the discursive soft fields (including law in this case), as Hyland (2002, p. 126) noted for research articles:

> The greater use of Discourse Act forms in the humanities and social sciences is more appropriate in an argument schema which more readily regards explicit interpretation, speculation, and complexity as accepted aspects of knowledge. [...] Writers in the soft disciplines therefore employed arguments that made greater use of Discourse Act forms which expedited the verbal exploration of such issues, facilitating qualitative arguments which rest on finely delineated interpretations and conceptualisations, rather than systematic scrutiny and precise measurement.

Through their posts, law bloggers try to gain credibility by establishing an appropriately "authorial persona" (Hyland, 2001). This is realised in the context of the argument that they construct. It is this argument that confers an authoritative discoursal identity on them, as shown in examples (4) and (5) taken from the corpus.

(4) *I would argue* that the case does not concern parliamentary sovereignty but parliamentary legislative supremacy. By this I mean that Parliament can legislate on any subject matter it wishes to save one it cannot bind its successors and that courts can interpret, but cannot contradict, clearly expressed statutory provisions. It does not entail that courts can only interpret legislation so as to uphold the intentions of Parliament. This is because *I argue* that parliamentary legislative supremacy is justified not because it aims to ensure that Parliament is supreme, but because it ensures that supremacy is shared between the legislature and the courts. (UKCLA)

(5) The President of the Supreme Court Lord Neuberger told 'The Times' last week that the idea is a 'recipe for complication, for cost and for

> unnecessary duplication', and the cross-bench peer Lord Pannick added that 'the proposal has no merit'. In this post, I am less kind to the idea. *I argue* that it is half-baked, unnecessary, and potentially dangerous, regardless of one's view of the desirability of an entrenched constitution protected by constitutional judicial review. (UKCLA)

As the examples show, the law blogger's role appears to be that of a writer who clearly assumes a role as arguer. Arguing about court decisions fuels the debate the blogger builds within his/her legal community. This result supports Kerr's (2006, pp. 1127–1128) view that writing blog posts allows law scholars to participate in debates on law-related topics and become "public intellectuals", but also to create a niche for subject-matter experts. As claimed by Tessuto (2015, p. 103), they promote "critical debate and knowledge about the law and legal institutions" and advance "legal scholarship".

Further evidence of the argumentative dimension of legal blogging comes from the analysis of the possessive *my*. It collocates with nouns such as *view* (6 occurrences), *mind* (3 occ.) that clearly indicate the adoption of the highly opinionated involvement of bloggers in discussing court decisions, as the following concordance lines show (Fig. 5.1):

As regards the object *me*, similar uses emerge in the corpus. When searching its collocates immediately following it and ranked by frequency (with a minimum frequency set at 2), the following three collocates appear on the list: *to me* (8 occurrences), *for me* (5 occ.), and *let* (1 occ.) as in *let me be clear*. No instances of the imperative *let me* followed by verbs of saying or verbs of the mental process type are found. *To me* and *for me* occur in initial position to preface a comment. Here are some concordance lines from the corpus (Fig. 5.2):

continue to be controversial. **In my view,** it is not for the
human rights demand. It is premised, **in my view,** on the following two
The present case affords, **to my mind,** a good example of the judic
are for a serious medical condition. **To my mind,** Coleman is noteworthy not
hose cases (who assert plausible, if **to my mind** flawed, free speech

Fig. 5.1 Concordances of *my* in law blogs (right sorted). (Source: SCOTUSblog, UKCLA)

```
yet still the Court granted review. To me that suggests a strong desire (by at
 and the statute was too uncertain. To me, the facts of these cases-and the
   be the key to unlocking Walker?  For me, the critical fact in the case is
```

Fig. 5.2 Concordances of *me* in law blogs (left sorted). (Source: SCOTUSblog)

```
ased on the oral argument, it seems to me that the Justices are closely
It seems like a very close call     to me but, if I had to guess, the
tarters, it is somewhat troubling   to me that a panel of the Commission
ot be sufficient but it was unclear to me whether any of the justices agreed
just upside down."It was not clear  to me of his take on the standing questio.
```

Fig. 5.3 Concordances of *me* in law blogs (left sorted). (Source: SCOTUSblog)

```
The oral arguments have convinced   me that a majority of the Court will
governor." Indeed, what strikes     me most in reading the brief is that
```

Fig. 5.4 Concordances of *me* in law blogs (left sorted). (Source: SCOTUSblog)

Another case is the *it* introductory pattern + *seems* or *is* + adjective + *to me* + extraposed complement clause (Fig. 5.3):

And also *me* as direct object in a clause (Fig. 5.4):

As the concordances show, law bloggers make themselves visible in their posts and present their own positions through commentaries expressing agreement and disagreement with court decisions, (i.e. *The present case affords, to my mind, a good example of…*; *it is somewhat troubling to me that a panel of the Commission found…*).

Collocational Profile of Subject and Object Pronouns *I* and *Me* and the Possessive *My* in Science Blogs

In the science blog posts, self-mention reflects the personal focus as well as the immediacy and spontaneity usually associated with the spoken medium, the top collocates including contracted forms, the cognitive *think* and other verbs having to do with mental processes (e.g. *doubt, suspect, find*), and a variety of attitudinal verbs (e.g. *fear, hope*), as shown in examples (6), (7), and (8):

(6) *I can't allow* this to go unchallenged. This research is so potentially harmful, and offers such little benefit to society, that *I fear* that NIH

is endangering the trust that Congress places in it. And don't misinterpret me: *I'm* a huge supporter of NIH, and *I've argued* before that it's one of the best investments the American public can make. But they got this one really, really wrong. (GEN_2019-03-04.txt)

(7) *I doubt* that "Lyme literate" doctors will accept the latest results and stop prescribing long-term antibiotic use; their websites indicate that they already know that they are right. *I hope*, though, that patients will start to question doctors who put them on long-term, possibly harmful antibiotic regimens that don't provide any benefit. (GEN_2016-04-04.txt)

(8) *I found* that suggestion a bit surprising. It certainly raises fascist images of GATACCA, even though the intention is completely benign. That aside, *I don't find* the suggestion very compelling. (NEU_2015-12-22.txt)

I is also frequently occurring in first-person narratives when the blogger tells a story or reports something personal that happened to him, often to frame a new scientific topic, as in example (9):

(9) Over the weekend, my wife and *I took a walk* with our son at the Storybrook Trail, an accessible trail with a fine overlook to the east over the Page Valley. (MB_2018-01-23.txt)

More specifically, the search for collocates of *I* returned 69 different collocates (ranked by significance and with minimum frequency set at 5). At the top of the list is the contracted form *I'm* and many other contractions appear among the most frequent collocates, namely, *I've, I'd, I'll.* This category of collocates points to the spokenness and informality of science blogs. The third most frequent collocate is the verb *suspect,* a cognitive verb that expresses the writer's critical stance.

If instead no minimum frequency requirement is set for collocates and all words co-occurring immediately to the right of *I* are counted, the 20 most frequent verb forms are *was, think, have, would, can, 'd, 'll, don't, found, am, took, saw, hope, had, and should*, with more contracted forms, a whole range of modal verbs, the link-verb *be* and some verbs of the mental process type such as *think, hope*, and *saw*. Of the mental processes, *I think* is the prototypical expression for marking subjective stance (as illustrated by the concordances in Fig. 5.5), making the writer's orientation explicit

```
   when Boyer eventually gets back to it. I think a good editor could have
    of the sites across the country, but  I think a stratigraphic column
       in front of such a large audience.  I think about how they'll feel 10 or
   their yawns.  What does all this mean? I think all this is best understood
        out to be much more powerful than I think anyone thought, virtually
ica. The second novel is far preferable,  I think because Jack and Toby
  cimals are covered by the fourth grade. I think Congress can handle
 .  The other thing Darrel alluded to but I think deserves to be expanded on
have in astonishing ways. Rocks can flow. I think folds show this most
    the timeline above, they'll know what I think: get rid of the whole lot,
    appeal to many of his supporters. But I think he represents an
      Johann Hari's article and see where I think he veers into denialism.
  sense of being in the volcanic deposit, I think Herculaneum was the more
 iology is not my professional specialty, I think I can make some general
the most prominent features of this blog. I think I find them so fascinating
my photo archives looking for folds that (I think)I haven't shared before,se
an especially good year for raptors, and  I think I'm getting better at 20
override the do-not-disturb mode, which   I think is a pretty terrible idea. I
and there is enough about it that         I think is applicable to the
Even the chaos theorist Ian Malcolm, who  I think is Crichton's most
sensory experience to trigger. Or (what   I think is far more likely) it
    in order to economize on space: This  I think, is instructive on two 25
me to assign poetry to students, but      I think it could be an option for
look roughly like this:    It went fine,  I think - it gave students a 27
ever accept the news of aspartame safety. I think it is a "business" 28
requency," and there's nothing to "slip." I think it's fair to say that
```

Fig. 5.5 Concordances of *I think* in science blogs (right sorted). (Source: GEN, MB, NEU)

while functioning like all other differentiations discussed in Sect. 3.3 (*I guess/imagine/suppose/etc.*, *I don't think, I thought*). *I saw* is used widely by the geology blogger who has many pictures of data he collected and for whom visualisation is an important communicative strategy, and *I hope* is also marking stance with an emotive connotation to it.

Similarly, collocates of *me* (ranked by statistical significance and without setting any minimum frequency) include a variety of verbs immediately preceding the pronoun such as *strikes/struck, reassured, pleases, impress, emailed, compelled, assures, amazes, worries/worry, pains, delights, encouraged, surprise, convince/convinced*, the majority of which belong to the category of affective/emotive verb types, delineating the attitudinal type of stance permeating science blogging.

If instead collocates to the left of *me* are calculated by considering only those collocates that occur a minimum of five times (Table 5.3), the most frequent combination that appears on top of the list is *Let me* occurring 20 times in all three blogs:

Table 5.3 Collocates of *me* in science blogs (1L and ranked by significance)

Rank	Freq (Left)	Statistic	Collocate
1	20	6.65118	let
2	8	6.61413	took
3	5	5.01972	including
4	5	4.96687	made
5	5	4.03596	get
6	7	3.81635	like
7	32	3.79090	for
8	17	3.41463	with
9	47	2.83516	to
10	5	2.04963	by

#Total no. of collocate types: 10
#Total no. of collocate tokens: 151

```
      hey are remarkably clear and accurate. Let me illustrate by revealing part
          .) While I've got your attention,  let me answer one of the top 10
                a lot to unpack here, but let me focus on a couple of
          in a one-paragraph review,     but let me say this: it is a compelling,
     as education, income, and self-control. Let me describe the study and then
    n cause sickness, disability, and death. Let me show you what Carson could
         a consequence of many small errors. Let me explain.  In almost all
                  of the 1970s-era SAT exams.  Let me put this another way: there's
        anti-vaxxers do this again.  Finally,  let me repeat something that can't
    gnosed with bowel cancer in 2012. First, let me say that of course I wish
               that.  How good is it? First, let me remind you that variant
     already packed. Fortunately, Con Gillen let me know about another spot
     summarize her points here, but instead  let me quote from one of the public
            rid of our reliance on integers. Let me explain. The root of our
        . It's not that complicated, really. Let me explain. EPA Administrator
    should we invest in biomedical research? Let me put some numbers on the table
   from this heartbreaking, nasty situation, let me say this: we have important
            cobbles are really hard to spot. Let me outline a few that I spotted,
    and bolster medical conspiracy theories. Let me dispense with the easy parts
    ower cost, than for-profit universities. Let me try another comparison.
```

Fig. 5.6 Concordances of *let me* in science blogs (left sorted). (Source: GEN, MB, NEU)

The expression points to the spoken conversational nature of blogs and to signposting language used by the bloggers to guide their readers through the argumentation flow (see *Let me illustrate, but let me focus on a couple of important points, Let me explain, Let me try another comparison* and the other instances in the concordance lines in Fig. 5.6). The verbs following the imperative *let me* are all verbs of saying and include *illustrate, answer, say, describe, explain, put, repeat, remind, quote, outline*, or verbs of the mental process type such as *focus on, show, dispense with, try*, showing the written argumentation function of blogs:

Another frequent collocate on this refined list is *including*, which, however, is used exclusively by Stephen Salzberg of *Genomics, Medicine and Pseudoscience* when referring to the scientific community to include himself in it. The collocate *like* has the same inclusive function as it occurs in the phrase *scientists like me*, used by Salzberg, but also by the other bloggers (*Educators like me*, *people like me*) showing that while affirming their own subjectivity, bloggers also position themselves as members of a community of practising scientists and educators. Interestingly, in two instances, *For those (like me) whose stomach is upset by aspirin...* (GEN) and *If like me you opt to pick it again...* (MB), the writer is reaching out to his readership and construing communal membership (see Myers, 2010, p. 4 on this function of blogging). The causative *made* also appears on the list of collocates, once again combined with verbs of the affective/emotive type (*made me wish, feel, wonder*). *Get* belongs in sequences *Don't get me wrong, don't get me started*, used by Steven Salzberg, who has the most conversational informal style of the three bloggers when addressing his audience; *for me* and *to me* reinforce the first-person perspective often appearing in parenthetical phrases expressing stance with the meaning of "in my opinion" (*This, to me, is the most interesting thing about...* or *it seems to me...*); *with me* appears twice in the expression *bear with me* used by two different bloggers to address the audience and finally *by me* occurs only in MB, pointing to an explicitly subjective style by Callan Bentley.

The search for collocates immediately following *my* with a minimum frequency set at 5 produced the following list of 17 collocates (ranked by frequency) (Table 5.4):

From the point of view of their semantic classification, the collocates show the frequent references to the personal sphere of the blogger (*wife, son, daughter, family*, and *friend(s)*) and to members of the blogger's professional circle (*colleague(s), students*). It should be noted that Callan Bentley of *Mountain Beltway* refers the most to his family environment and uses a narrative style. Another group is *opinion, own, mind*, and *experience*, which belong in the fixed expressions *in my opinion*, used

Table 5.4 Collocates of *my* in science blogs (1R ranked by frequency)

Rank	Freq (Left)	Statistic	Collocate
1	24	6.56510	own
2	14	8.56510	wife
3	14	8.20253	son
4	12	8.48710	favorite
5	9	7.27559	family
6	9	8.23252	colleague
7	8	6.47764	mind
8	8	8.34271	friend
9	7	8.63549	yard
10	7	4.00905	first
11	7	6.02668	colleagues
12	6	7.21146	opinion
13	6	6.18044	experience
14	6	7.48710	daughter
15	5	4.78549	students
16	5	7.07967	friends
17	5	4.96419	field

#Total no. of collocate types: 17
#Total no. of collocate tokens: 152

exclusively by one blogger, Steven Novella, and *in my mind, to my mind, in my experience*, used in *Mountain Beltway*. Together with the intensifier *own*, all the collocates in this group form expressions used for emphasising the speaker's expressing a personal opinion about something.

The remaining *yard*, on top of the list, *field*, and *first* at the bottom of the list, are both used by the geology blogger of *Mountain Beltway*, who writes that he keeps a "yard list" of all the bird species he observes in his yard, advises about the habit of keeping *field notebooks* and refers to his *first* day of field work. This could be interpreted as one way of defining the physical identity of the blogger through indications of place, what Myers (2010) calls "place pins one down" as opposed to the placelessness of the blogosphere. According to Myers, place references act rhetorically in blogs, for example, to justify a statement or introduce a narrative (2010, p. 51). By including the physical reference to his yard, the blogger tells something specific about himself to start telling a story (story-telling function), ultimately revealing something of himself. Together with the

```
   Galapagos. The previous week (over Christmas) my family and I were in
      serious and scary disease. It has touched  my family, and I am sure
thing completely different!  This past weekend, my family gathered in Capon
             's still a good idea. Everyone in  my family got it.)  Oh, and
            .  ...And so: this image is going on  my family's Christmas card
    of interesting details. Hiking over it with  my family was the geological
   old lighthouse. I went there yesterday with   my family. We're on vacation
```

Fig. 5.7 Concordances of *my family* (right-sorted) in science blogs. (Source: GEN, MB, NEU)

references to family members and vacations or activities done together, they all help to "celebrate one's routine everyday world" (Myers, 2010, p. 55). In general *Mountain Beltway* is the blog where most of the collocates of *my* occur, as can be seen in the concordance lines in Fig. 5.7 of *my family*, almost all from his blog. This might indicate a stylistic peculiarity of Callan Bentley, while it also recalls what Bondi (2022, p. 8) has called the "biographical individual self" in relation to the same blogs.

3.3 Functional Analysis of Stance

The verbs co-occurring with the subject pronoun *I* and object pronoun *me* can be classified into semantic categories of process types as in Halliday and Matthiessen's classification (2014, pp. 213–216). These include material, mental, relational, verbal, behavioural, and existential. As can be noticed in Table 5.5, the process type that is most represented in both the law and science blogs is the mental process type, which can be further broken down into its four different subtypes to see which has the highest number of verb types in the blogs considered. The four subtypes are the perceptive, cognitive, desiderative, and emotive (Halliday & Matthiessen, 2014, pp. 256–257). The variety of verbs that combine with both the subject and object first-person roles testify to the centrality of the semantic role of Senser, that is, the one who senses, feels, thinks, or perceives (ibid., p. 249), which construes the blogger, with their personal attitudes and opinions, as the undisputed protagonist of the writing.

The other process types that are prominent are (in order of frequency) the verbal process type realised by verbs of saying. The material process type (verbs of doing), when, for example, the bloggers recount what they

Table 5.5 Representation of blogger's identity in the law and science blogs

Process types	Verbs in science blogs	Verbs in law blogs
Verbs of affection (emotive)		
Subj. + verb	*I particularly delighted in, I was delighted to; I liked; I'm glad; I regret; I appreciate; I'm grateful to/that; I love/loved, I would love; I'm/was eager; I enjoyed; I welcome; I missed; I fear; I feel/felt; I was curious; I was surprised, I am aghast; I am flabbergasted*	–
Verb + obj.	*affects me; amazes me; delights me; emboldened me; encouraged me; impress me; inspired me; motivate me; pains me; pleases me; surprise me; worries/worry me*	–
Verbs of cognition (cognitive)		
Subj. + verb	*I suspect, I guess, I'm guessing, I figured, I think, I thought, wonder, I appreciate, believe, consider, miss, I realized, I doubt, I knew, I suppose, I know*	*I believe, I think, I thought, I don't think, I doubt, I know, I understand, I'm guessing, I consider, I found, I focus on, I suppose, I see, I do not see, I analyze*
Verb + obj.	*compelled me; convinced me; lost me; lured me in; reminded/reminds me; strikes/struck me*	*strikes me*
Verbs of volition (desiderative)		
Subj. + verb	*I intend, I wanted, I hope, I wish, I agree*	*I intend, I would like, I would have liked, I agree*
Verb + obj.	–	–
Verbs of perception (perceptive)		
Subj. + verb	*I've seen, I've spotted*	–
Verbs of saying		
Subj. + verb	*I argue, I asked, I digress, I checked, I talked, I blogged, I wrote, I mentioned, I note, I highlighted, I admit, I've pointed out, I posted, I quote, I suggest, I urged*	*I argue, I argued, I have argued, I will argue, I would argue, I have concluded, I mentioned, I suggest, I have suggested, I do not suggest, I wrote, I note, I observed*
Verbs of doing		
Subj. + verb	*I encountered, I finished, I started, I picked, I spent, I took a trip/a hike, I've swum, I visited, I've waded*	–

did with their family, is limited to the science blogs. Here the blogger is semantically represented as Actor engaging in some more or less mundane activity as part of a narrative section in the blog (see Myers, 2010 and Luzón, 2012 on the role of narratives in academic blogs).

The verbs of saying have an important function to signal the writer's argumentation and to indicate the blogger's stance, as can be seen, for instance, in (10) from the law blogs and (11) from the science blogs.

(10) [...] Based on the paucity of commentary, it seems that few Supreme Court watchers or legal commentators followed the case. It appeared to present a rather narrow statutory interpretation question about the eligibility of a small group of children (children conceived through the use of assisted reproductive technology after the death of the sperm provider) to a particular federal benefit. *I argue*, however, that the Court's decision in this case may provide useful insights to how the Court may respond to some much broader and controversial arguments currently being asserted in the various same-sex marriage cases percolating through the court system. (SCOTUSblog)

(11) Why, then, are there frequent headlines saying that a study shows that acupuncture works for this or that? That's a good question. We are far past the point with acupuncture that doing preliminary studies is of any value. *I would argue* there is no point to even rigorous studies—three thousand studies is enough already. But if you are going to do a study of acupuncture, if it is less than rigorous it is completely worthless. Even worse, it is likely to be misleading. The only point would be to confirm what you want to believe, and to generate pro-acupuncture headlines. (NEU_2017-01-23.txt)

According to Hyland (2005b, pp. 149–151), writer's attitude is "most explicitly signalled by attitude verbs (e.g., *agree, prefer*), sentence adverbs (*unfortunately, hopefully*) and adjectives (*appropriate, logical, remarkable*)" in research articles in the humanities and social sciences rather than in the sciences, "where they contribute to a writer's persona and establish a link with the disciplinary community" (Hyland, 2005b, p. 149). However, mental verbs of affection and volition (the emotive and desiderative types in Table 5.5) are over-represented in the science blogs, making attitudinal stance prominent in this kind of blogging, compared to the law blogs where only some verbs of volition are used. The findings so far might adumbrate two different subgenres of research blogs (see Luzón & Pérez-Llantada, 2019), as we discuss in the next section.

4 Discussion and Conclusions

The analysis of the frequency distribution and collocational profile of first-person pronouns *I/me* and of the possessive *my* has allowed us to explore the bloggers' construal of identity and the different types of stance realised in the law and science blogs. When comparing the two types of blogs, the analysis has revealed two different foci: while law blogs focus on the technical contents (revealed by the lower frequency of *I/me* in law compared to science), science blogs mark subjective assessment explicitly and through a variety of attitudinal verbs. This difference reflects the different roles of the bloggers in the two samples considered and the different communicative functions of the blogs. If law bloggers comment on a judge's or a court's decision on legal cases with the purpose of introducing context of knowledge discussion within the discipline, the conversation started by the blog being mainly ideational, science bloggers vent their opinions on science matters to engage in a public debate, thus making the interpersonal dimension prominent.

When zooming in on expressions of stance, we noticed that some are common to both blogs (the opinion marker *I think*, the prepositional/adverbials *for me/to me, in/to my mind, in my opinion* used to preface a comment). However, a more fine-grained analysis of verbs co-occurring with *I/me* revealed that in the law blogs, the preferred verb is *argue*, while a whole set of attitudinal verbs of the mental affective/emotive categories are used exclusively in the science blogs. A category of mental process types shared by the two blogs is the desiderative, exemplified by *intend, like*, and *agree*. This testifies to the function of blogs as platforms for self-disclosure, the blurring of boundaries between private and public spheres and community building, as already shown by Myers (2010) and

Mauranen (2013) and illustrated by the frequency of the desiderative and attitudinal verb *agree*. As for the verbs of saying, *argue* reflects the argumentative purpose of law blogs, while in science blogs it appears together with other verbs of saying that the bloggers use not only to develop their reasoning (e.g. *I would argue*) and express their opinion, but also to engage with their readers through a variety of speech acts (e.g. *I'd recommend it…; I strongly encourage you to…; I urge you to…*). The verbs of saying that they have in common (e.g. *write, mention, note, suggest*) reflect the medium the bloggers are using (e.g. *write, blog, post*) and the writer's explicit marking of the textual development (e.g. *mention, note, highlight, observe, suggest*), thus identifying bloggers as rhetors who deploy textual markers to "create a particular pragmatic or discoursal effect" (as in Hyland, 2005b, p. 28).

Law bloggers under investigation write their posts to discuss legal issues "whose value simply turns on the rigor of the analysis in the underlying ideas they express" (Peoples, 2010, p. 51). Science bloggers instead use blogs to express their opinions and personal experiences freely, thus as a form of self-disclosure and self-expression closer in communicative function to the early personal blog genre (Myers, 2010) as well as to engage with the scientific community and a broader public, create a forum for discussion and ultimately promote public understanding of science.

In conclusion, the blogs examined reflect specialisation related to their different communicative functions. On the one hand, the law blogs have argumentation construction within the legal community as a dominant feature of authorial stance and identity. On the other hand, while also revealing a common argumentative function of blogs, the science blogs have self-expression and public engagement in science as overarching communicative purposes. Further investigations are necessary to explore if such functional specialisation can evolve and stabilise into two different genres or subgenres of research writing in the blogosphere.

Acknowledgements This research was supported by the Italian Ministry of Research, grant number PRIN 2015 2015TJ8ZAS.

References

Anthony, L. (2022). *AntConc* (Version 4.1.4) [Computer Software]. Waseda University.

Bondi, M. (2022). Dialogicity in individual and institutional scientific blogs. *Publications, 10*(9), 1–20. https://doi.org/10.3390/publications10010009

Diani, G. (2021). "In this post, I argue that…": Constructing argumentative discourse in scholarly law blog posts. *European Journal of English Studies, 25*(3), 369–384. https://doi.org/10.1080/13825577.2021.1988256

Diani, G. (2022). Managing discussions in law blogs: From post to comments. *International Journal of Law, Language & Discourse, 10*(2), 9–21. https://doi.org/10.56498/1022022409

Freddi, M. (2020). Blurring the lines between genres and audiences: Interaction in science blogs. *Discourse and Interaction, 13*(2), 9–35. https://doi.org/10.5817/DI2020-2-9

Garzone, G. (2014). Investigating blawgs through corpus linguistics: Issues of generic integrity. In M. Gotti & D. S. Giannoni (Eds.), *Corpus analysis for descriptive and pedagogical purposes: ESP perspectives* (pp. 167–188). Peter Lang.

Halliday, M. A. K., & Matthiessen, C. M. I. M. (2014). *Halliday's introduction to functional grammar*. Routledge.

Hyland, K. (2000). *Disciplinary discourses: Social interactions in academic writing*. Longman.

Hyland, K. (2001). Humble servants of the discipline? Self-mention in research articles. *English for Specific Purposes, 20*, 207–226. https://doi.org/10.1016/S0889-4906(00)00012-0

Hyland, K. (2002). Activity and evaluation: Reporting practices in academic writing. In J. Flowerdew (Ed.), *Academic discourse* (pp. 115–130). Longman.

Hyland, K. (2005a). Stance and engagement: A model of interaction in academic discourse. *Discourse Studies, 7*(2), 173–192. https://doi.org/10.1177/1461445605050365

Hyland, K. (2005b). *Metadiscourse: Exploring interaction in writing*. Bloomsbury (Continuum).

Hyland, K., & Jiang, F. (2019). *Academic discourse and global publishing. Disciplinary persuasion in changing times*. Routledge.

Hyland, K., & Zou, H. (2021). "I believe the findings are fascinating": Stance in three-minute theses. *Journal of English for Academic Purposes, 50*(2), 1–13. https://doi.org/10.1016/j.jeap.2021.100973

Kerr, O. S. (2006). Blogs and legal academy. *Washington University Law Review, 84*(5), 1127–1134.

Luzón, M. J. (2011). "Interesting post, but I disagree": Social presence and anti-social behaviour in academic weblogs. *Applied Linguistics, 32*(5), 517–540. https://doi.org/10.1093/applin/amr021

Luzón, M. J. (2012). Narratives in academic blogs. In M. Gotti & C. Sancho Guinda (Eds.), *Narratives in academic and professional genres* (pp. 175–193). Peter Lang.

Luzón, M. J. (2013). "This is an erroneous argument": Conflict in academic blog discussions. *Discourse, Context & Media, 2*, 111–119. https://doi.org/10.1016/j.dcm.2013.04.005

Luzón, M. J., & Pérez-Llantada, C. (2019). *Science communication on the internet. Old genres meet new genres.* Benjamins.

Mauranen, A. (2013). Hybridism, edutainment, and doubt: Science blogging finding its feet. *Nordic Journal of English Studies, 13*(1), 7–36. https://doi.org/10.35360/njes.274

Myers, G. (2010). *The discourse of blogs and wikis.* Routledge.

Peoples, L. F. (2010). The citation of blogs in judicial opinions. *Tulane Journal of Technology & Intellectual Property, 13*, 39–80.

Tessuto, G. (2015). "Posted by…": Scholarly legal blogs as part of academic discourse and site for stance and engagement. *Textus, 28*(2), 85–107.

Zou, H., & Hyland, K. (2020). Academic blogging: Writers' views on interacting with readers. *Ibérica, 39*, 267–294. http://www.revistaiberica.org/index.php/iberica/article/view/84

Zou, H., & Hyland, K. (2022). Stance in academic blogs and three-minute theses. *International Journal of Applied Linguistics, 32*(2), 225–240. https://doi.org/10.1111/ijal.12411

Part III

Visibility and Dissemination in Scientific Research Contexts

6

Predatory Journals: A Potential Threat to the Dissemination of Open Access Knowledge

Ana Bocanegra-Valle

1 Introduction

The pressure-to-publish trend in academia has contributed to the proliferation and expansion of journals from predatory publishers (Fazel & Hartse, 2020; Forero et al., 2018; Kurt, 2018). Many of these publishers usually email scholars promising short peer review turnarounds, high acceptance rates, and fast publication (Beall, 2016; Bocanegra-Valle, 2017; Soler & Cooper, 2019). In the extant literature, these journals are called *predatory* because, although they claim to be legitimate, they prey on unguarded scholars (Laine & Winker, 2017), misrepresent current academic publishing practices (Elmore & Weston, 2020) and exploit the author-pays model (gold open access) for their own profit with no intention of providing the expected editorial services in return (Beall, 2012; Fazel & Hartse, 2020; Shamseer et al., 2017). By the beginning of 2017, Beall's "List of predatory publishers" contained 1294 titles of potential,

A. Bocanegra-Valle (✉)
Universidad de Cádiz, Cádiz, Spain
e-mail: ana.bocanegra@uca.es

R. Plo-Alastrué, I. Corona (eds.), *Digital Scientific Communication*,
https://doi.org/10.1007/978-3-031-38207-9_6

possible, or probably predatory scholarly open access journals, showing a sharp increase from 18 journals in 2011.

Researchers within the health sciences are particularly concerned with predatory practices and how to identify them (e.g. Beall, 2013, 2016; Cobey et al., 2018; Elmore & Weston, 2020; Forero et al., 2018; Laine & Winker, 2017; Memon, 2019; Shamseer et al., 2017); even websites have been set up to help researchers identify trusted journals (e.g. thinkchecksubmit.org). In other disciplines, like Applied Linguistics, information for scholars and researchers is certainly very scant. Some notable exceptions are Fazel and Hartse (2018), who investigated the most notable features of twenty "potentially predatory" journals; Soler and Cooper (2019), who analysed the rhetorical moves of contact emails from predatory journals; Soler and Wang (2019), who compared the linguistic differences between a well-established and a predatory journal in political science; Fazel and Hartse (2020), who discussed the implications of predatory publishing for the field of English for research publication purposes (ERPP); Pecorari (2021), who examined the main features of predatory conferences; and Pflugfelder (2022), who explored the rhetorical features of predatory articles. More recently, this author has also investigated the interpersonal resources (like engagement markers) used by predatory journal editors to persuade prospective authors (Bocanegra-Valle, 2023).

This study investigates the narrative that predatory journal editors employ to appear trustworthy and generate interest among prospective authors. With a particular focus on predatory journals in the field of Applied Linguistics, this study pursues the general aims of raising awareness among applied linguists as well as providing guidance and educating prospective authors (particularly junior researchers) in recognising predatory journals. It also critically assesses the ways predatory publishing practices may impact on authors' academic career, the dissemination of knowledge, and the expansion of open access practices in scholarly communication.

2 Predatory Publishing Practices

The expansion of digital editorial practices and the global trend towards the open dissemination of science has contributed to the flourishing of academic journals around the world. In this context, some publishers email flattering paper requests to scholars that offer prompt peer reviews and fast open access publication. These unsolicited invitations, known as *spam emails* (Beall, 2015; Kurt, 2018) or *fishing emails* (Gasparyan et al., 2015), are among one of the most frequent strategies that predatory publishers employ and, from their very reception, should alert researchers about illegitimate publication practices (Forero et al., 2018).

As opposed to *legitimate* or *trusted* journals, the names of predatory journals abound in current literature: *illegitimate journals*, *counterfeit journals*, *pseudo-journals*, *pseudoscientific journals*, *fraudulent journals*, *questionable journals*, *sub-par journals*, *vanity journals*, *parasitic journals*, *spam journals*, *deceptive journals*, *scamming journals*, *fake journals*, *paradoxical journals*, *parodical journals*, *fast-turnaround journals*, *spoofy journals*, *dodgy journals*, *dark journals*, *sham journals*, or *journals operating in bad faith*.

Regardless of this diversity of labels, predatory journals share the same business model: on the one hand, researchers pay to have their article published and made freely available on the internet; on the other, publishers exploit the author-pays model (gold open access model) for their own profit, without providing the editorial and publishing services associated with legitimate journals—namely, quality control through editorial and peer review, licensing, or indexing. In most cases, authors are unaware of the fact that they have to pay until they are surprised by an invoice from the publisher (Beall, 2013, 2015, 2016; Laine & Winker, 2017). Then, if authors do not want to pay the article publication charges and ask for the article to be withdrawn, they are required to pay a *withdrawal fee*. Because legitimate journals request a submission declaration from authors to assure that the work is original, that has not been published previously or that is not under consideration for publication elsewhere, by not accepting to pay this fee, authors are prevented from submitting their work to other journals, and therefore this article is held *hostage* until a *ransom* is paid (Beall, 2016). Another interesting concept within these security-related metaphors employed in the current literature to refer to predatory publishing practices is that of *highjacked* journals. These are journals that create a fake website that resembles the website of a trusted journal "for the purpose of soliciting submissions and collecting author fees from authors who believe they are sending their work to the legitimate journal" (Laine & Winker, 2017, p. 286).

The spread of predatory practices, particularly journals, is due to three main, and at the same time interrelated, factors that are well documented in the literature (e.g. Fazel & Hartse, 2020; Kurt, 2018; Laine & Winker, 2017; Memon, 2019; Shamseer et al., 2017; Xia et al., 2015): the pressure to "publish or perish," the rise of open access, and the availability of electronic publishing.

In the first place, the publish-or-perish pressure to publish in English language, internationally renowned journals, prompts researchers and scholars to look for rapid publication, with high acceptance rates and quick peer reviews. This may seem an attractive and straightforward option for many researchers, particularly doctoral candidates and early-career researchers, who need to demonstrate publication outcomes. It may also be the case for both inexperienced and experienced researchers from emerging countries who need to be positively assessed in order to earn tenure or promotion, or are compelled to position their institution in prestigious rankings. Against this background, deceptive journals thrive and prosper because they provide young researchers and scholars with "a low barrier to publication" (Laine & Winker, 2017, p. 289).

The second factor is the rise of open access publishing, which provides scholars with the opportunities to disseminate their work easily, quickly, and globally. "Open access" is a buzzword in predatory-related bibliography, and it is widely acknowledged therein that even though open access does not lie at the origin of fraudulent practices, it has created the conditions for predatory publishing practices to flourish and aggravated other problems related to open science. By way of example, it "impinges on the reputation of certain journals, particularly those accessed or published exclusively online; young, small or peripheral journals (…); or those that are not (or not yet) abstracted or indexed in reputable databases and lists" (Bocanegra-Valle, 2017, pp. 121–122).

Thirdly, the growth of electronic publishing provides a broad and accessible platform for publication at the same time that handling and production costs are reduced. On line journals and websites that mimic legitimate journal websites can be easily launched from home with a good computer, internet connection, and some editorial skills. Thus, in this favourable context, journals emerge, flourish, and disappear rapidly and with impunity, creating "short-cuts for illegitimate publishers" (Gasparyan et al., 2015, p. 1014) and cheating authors who might have raised high expectations about their work.

Predatory journals should be clearly differentiated from *low-quality* journals. Whereas profit is the only driver for predatory journals, thereby flouting the expected ethical standards and editorial services from legitimate scientific publishers, low-quality journals may apply article processing charges simply because their infrastructure is poor; they are not supported by robust publishers and, thus, funding is insufficient; they might not (yet) be indexed in internationally-recognised databases; or they may face other problems like limited distribution or lack of readership. However, they implement rigorous peer review processes, abide by ethical codes of conduct, and work with sufficiently transparent editorial standards that allow for their trust—for specific criteria that differentiate deceptive journals, low-quality legitimate journals, and high-quality open access journals see Memon (2019).

3 The Study

3.1 Research Goals

This chapter aims at raising awareness of predatory publishing practices in the field of Applied Linguistics and at educating researchers and academics about the ways predatory publishing practices impact on authors' academic career, the dissemination of knowledge, and the growth of open access publishing practices. In particular, it sets out to (1) investigate the most salient themes that predatory email messages contain so as to appeal readers and generate interest for publication; (2) shed light on deceptive editorial practices that may threaten academic publication standards; and (3) reflect on the contribution of predatory publishers to the dissemination of non-validated knowledge and the ways these practices affect open access practices and scholars throughout the world.

3.2 Data Collection and Analysis

The investigation draws on a corpus of 50 email messages sent by predatory publishers and directly addressed to the author of this chapter. Seventy email messages were collected during the past two years, and 50 were found to be representative and selected for the corpus. The reasons for rejecting the other messages were three: same message sent from a different e-mail address, same message sent for another journal title, or messages containing cryptic or illegible text. This dataset was qualitatively analysed by means of NVivo 11 Pro (Bocanegra-Valle, 2020). At this stage, three extremely helpful sources were: (1) Beall's (2015) criteria for determining predatory scholarly open access journals and publishers; (2) Cobey et al.'s (2018) characteristics of predatory journals in the health sciences; and (3) the interactive website https://thinkchecksubmit.org/, which helps scholars in the identification of legitimate journals that can be trusted for research publication.

First, the texts were scanned for key issues as documented in the literature (see Sects. 1 and 2 above). Then, these issues were organised into categories at the same time that three main themes were identified. Next, detailed coding was conducted—that is, broad categories were coded into initial nodes per theme, and later combined and reorganised to yield tree and concept maps that could help to visually analyse results. After this, the categories were thematically grouped as first- and second-level categories; hence, establishing a relational hierarchy. Lastly, samples were extracted and text queries for key words and collocations were conducted in order to refine the analysis. Table 6.1 shows the three main themes and categories (that is, "nodes" according to NVivo) that were identified together with the total number of occurrences (that is, "references") that are registered, and the total number of sources (e-mail message in the corpus) in which they were found.

Table 6.1 Main categories (node), sources, and occurrences (references) provided by NVivo

Node	Sources	References
Journal information	44	169
Counterfeit titles	10	10
Editors	26	38
Contact email	6	6
Location	16	16
Faulty language	10	12
Publication process	33	81
Cost	11	14
Prompt and fast	25	34
Publication frequency	11	11
Publication modes	6	6
Publication times	3	3
Submission	12	13
Unusual editorial practices	17	28
Several journals	6	7
Boastful language	35	97
Author recruitment	4	5
Flattering scholar	23	35
Kind phrasing	14	25
Self-reference	15	21
Unusual terminology	5	11
Journal quality	44	151
Bibliometrics	30	60
Copyright	3	6
Deceptive references	16	20
Impact factor	8	10
Indexing	15	17
Other indicators	7	7
Internationalisation	16	22
Open access	13	18
Originality	9	9
Peer review	25	27
Reputed affiliations	13	15

4 Results and Discussion

The coding results revealed three main themes, "Journal Information," "Boastful Language," and "Journal Quality," together with a number of first- and second-level categories per theme. These are presented and discussed in the next subsections.

4.1 Theme 1: Journal Information

Table 6.2 summarises the first- and second-level categories for the first theme "Journal Information." These mainly include journal titles (1.1), editors' information (1.2a-b), use of faulty language (1.3), unusual editorial practices (1.4, 1.4a), and other issues related to the publication process (1.5a-f).

From Table 6.2, we can deduce that in most predatory journals, titles mimic popular titles in legitimate journals (like the *Modern Language Journal* or the *Journal of Language Teaching Research*—see Excerpt 1.1). In an attempt to misrepresent their true location, predatory publishers do not usually state their headquarters location (Beall, 2015, 2016), or contact details are non-institutional addresses, always from free providers like gmail or yahoo (Excerpt 1.2a). In agreement with the literature (e.g. Beall, 2012; Byard, 2016; Kurt, 2018; Shamseer et al., 2017), the data analysed showed that these journals are mostly advertised from low-resourced countries like India, Nigeria, Iran, Turkey, Malaysia, or Pakistan—nothing to do with countries in which international well-established journals are often located (Excerpt 1.2b). These may suggest that low-resourced countries are breeding grounds for predatory publishers. Many spam emails are riddled with typographical, spelling, or grammatical errors, or use of inappropriate terms (see category 1.3). They also exhibit unusual editorial practices: the issuance of certificates as a proof of publication (Excerpt 1.4); the publication of either a vast number of research areas under the same journal (journals are very broad) or different journals in a variety of disciplines by the same publishers (Excerpt 1.4a). In agreement with Beall (2013, p. 48) and Laine and Winker (2017, p. 289), the publication process also contains a number of "warning signs" or "red flags" (see also, Fazel & Hartse, 2020; Gasparyan et al., 2015). These red flags are illustrated in excerpts 1.5a–1.5f of the corpus under study and mainly refer to article processing charges, short turnarounds, and lack of archived content:

- Article processing charges (APCs): APCs are all too often not clearly stated in the message, thereby lacking transparency. An author will

Table 6.2 Categories and examples for Theme 1 "Journal Information"

Theme 1: Journal Information		
Category		Email excerpts
1.1	Counterfeit titles	Modern Journal of [...] (E41)[a]
1.2a	Editors: contact email	Email: editor.[journal abbreviation]@gmail.com (E4)
1.2b	Editors: location	[correspondence address], Malaysia (E13)
1.3	Faulty language	See typos marked with an * in excerpts here and in subsequent tables
1.4	Unusual editorial practices	We are glad to inform you, now you will receive One Hard Copy of the Journal and Certificates for all the Authors (E14)
1.4a	Unusual editorial practices: several journals	Journal for All Subjects (E5)
1.5a	Publication process: cost	It's requested the contribution* of the authors with the fee of $45 USD per article to be published. This amount, which can be paid through PayPal or bank deposit/transfer, is allocated to the costs of diagramming, formatting and maintenance of the magazine* (E47)
1.5b	Publication process: prompt and fast	All manuscripts are peer reviewed and reviewing process will be completed in one week (E19)
1.5c	Publication process: frequency	12 issues for* year (E1)
1.5d	Publication process: modes	Online and print versions (e-journal and print) with reasonable cost (E1)
1.5e	Publication process: times	Manuscript Submission: 25 April 2021 Acknowledgement Notification: within 24 hours. Acceptance Notification: Paper publication: within 10 days 05 May 2021 (E70)
1.5f	Publication process: submission	The Journal accepts article submissions online or by e-mail (E13)

[a]Code in brackets refers to the message in the corpus. To safeguard anonymity, journal details which might lead to journal or publishers' identification have been removed in some excerpts.

become aware of the existence of APCs after submitting an article or if consulting the journal website (which may not exist, be inactive, or post the price in a concealed section). The charges may be disguised as open access fees, processing fees, operating costs, handling costs or editorial costs. It should be noted at this point that applying APCs alone does not qualify a journal as predatory, because small or poorly funded journals may need to defray editorial costs—as discussed above in Sect. 2 with respect to low-quality journals. It is this key point of APCs, together with deceptive editorial practices (like lack of proper peer review or ethical standards—see publication process and journal quality categories in the next subsection) which contribute to the making of illegitimate journals.

- Short turnarounds: the times and frequency of publication clearly indicate that peer review is much shorter than in legitimate journals. Journals promise rapid turnaround times (even 24 hours) from submission to publication, review included, and so they "make their peer review processes suspect if not impossible" (Kurt, 2018, p. 142). Short publication times have been found to be a key motivator when deciding to publish in predatory journals (Fazel & Hartse, 2020), but it is also a warning sign that should make prospective authors feel, at least, suspicious about journal legitimacy.
- Lack of archived content: published content is not archived in third-party sites and, because journals emerge and disappear rapidly, content may also become inaccessible in the future (Ferris & Winker, 2017). Articles are submitted by sending the manuscript to an email address and with no clear guidelines for authors. These journals do not usually process articles via editorial management software, which would help to exert greater control, minimise errors, improve the overall management quality of the journal, and safeguard perpetual content preservation (Bocanegra-Valle, 2017).

4.2 Theme 2: Boastful Language

Table 6.3 contains the categories for the second theme "Boastful Language". These refer to the use of persuasive language with a view to eventual article submission.

Excerpts in Table 6.3 reveal that predatory journals make use of recurrent flattery in their fishing emails. By praising authors and using direct and friendly language (see italics in Excerpts 2.3, 2.4 & 2.5), predatory publishers aim at highlighting their academic standing and constructing email recipients as experts or opinion leaders in their respective fields (Soler & Cooper, 2019). This boastful language also applies to the journals themselves which are often presented as outstanding, rigorous, prestigious, highly scientific, or leading in their fields (Excerpts 2.1 & 2.2). As Shamseer et al. (2017, p. 2) observe, such attractive invitations may be "an easy solution to inexperienced or early career researchers who need to publish in order to advance their career, or to those desperate to get a publication accepted after a number of rejections, or to those simply not paying attention". For others, this use of language may appear aggressive and therefore lead to journal questioning and scepticism (Xia et al., 2015).

Table 6.3 Categories and examples for Theme 2 "Boastful Language"

Theme 2: Boastful Language	
Category	**Email excerpts**
2.1 Unusual terminology	The perception of the journal is to bequeath with* an academic podium to researchers across the global* to publish their original, innovative, pragmatic and high-quality research work (E14)
2.2 Self-reference	It provides a forum for the publication of the most-advanced and rigorous scientific research related to... (E9)
2.3 Kind phrasing	My name is [...], and I am the editorial assistant of [...] I have had an opportunity to read your paper entitled [...] *and can tell from you work that you are an expert in this field* (E2)
2.4 Flattering scholar	Your article under the title of [...], which has been published in [...], *has left us a deep impression* (E45)
2.5 Author recruitment	It is privileged* for us to invite you to join our team as the editorial board member or reviewer. *We really hope you can be a part of our team* (E45)

Besides encouraging article submission, predatory emails often invite authors to join their editorial board, become reviewers on a permanent basis or edit special issues (Excerpt 2.5). In the case of all these boastful language categories, the overuse of first-person plural pronouns and adjectives helps to build a feeling of community which may appeal to authors, particularly inexperienced researchers and those "on the fringes of their communities of practice" (Bloch, 2021, p. 194)—that is, researchers working in isolation outside research groups and networks, or scholars from peripheral scientific countries (see a more in-depth discussion about this in Sect. 5.3).

4.3 Theme 3: Journal Quality

Table 6.4 contains the first- and second-level categories for the third theme "Journal Quality." These refer mainly to issues of internationalisation (3.1), open access (3.2), originality (3.3), affiliation (3.4), peer review (3.5), and bibliometrics (3.6a-e).

Excerpts in Table 6.4 indicate that issues of internationalisation, open access, and originality are associated with the quality of a journal and perceived as positive indicators of journal trustworthiness (Excerpts 3.1 to 3.3). References to the international scope of the journal address researchers' need for a global dissemination of their work. Open access is also an attractive option for researchers to make their work more visible and more citable—a clear driver in the case of national journals (Bocanegra-Valle, 2019). Lastly, requests for originality help to make the submission process legitimate and more credible.

In line with many international and national journals (see Bocanegra-Valle, 2019), journals claim to be affiliated with universities, established scholarly societies, academic institutions, or reputed organisations (Excerpt 3.4). Nevertheless, these may be fake or do exist but have not given consent to such supposed affiliation (Shamseer et al., 2017). Importantly enough, literature has attested that some of these journals list renowned scholars on their editorial boards without their knowledge or consent (Byard, 2016; Laine & Winker, 2017). Journal affiliations also contain the adjectives European, American, or British to appear more

Table 6.4 Categories and examples for Theme 3 "Journal Quality"

Theme 3: Journal Quality		
Category		Email excerpts
3.1	Internationalisation	It will *be allow you to have millions of people worldwide reading your work (E20)
3.2	Open access	Adhere to the open access mode, we report the latest scientific achievements and academic trends in the relevant areas (E18)
3.3	Originality	It intended* to publish original Research Papers and Review Papers (E28)
3.4	Reputed affiliations	Our journal in association with American Society of Scientific Research (E8)
3.5	Peer review	Papers for publication in the International Journal of [...] are selected through precise peer-review to ensure quality, originality, appropriateness, significance and readability (E39)
3.6a	Bibliometrics: copyright	Copyrights for articles are retained by the authors, with first publication rights granted to the journal. The journal/publisher is not responsible for subsequent uses of the work (E24)
3.6b	Bibliometrics: deceptive references	All the papers shall be uploaded on HarvardDataverse, Google Scholar, Academia and *Ochid (E3)
3.6c	Bibliometrics: impact factor	Impact Factor:1.17; h5-index: 14; Downloads> 1,174,471; Published Articles> 371; Citations> 1635; Downloads/article: 3165.7; Citations/ article:4.4 (E50)
3.6d	Bibliometrics: indexing	Indexed and abstracted in Chemical Abstracts, Index Copernicus, DOAJ, Open J-Gate, NewJour, Science Central, and Google Scholar (E19)
3.6e	Bibliometrics: other indicators	[...] Journals provides* DOI (Digital Object Identifier) to each article (E40)

trustworthy (even though the journal's headquarters are located in other continents). It may also be the case that journals are located "in odd places such as deserts and garages" (Bloch, 2021, p. 194).

Peer review is the hallmark for quality in scholarly publishing and, therefore, the most contentious attribute in predatory practices discussions (e.g. Byard, 2016; Fazel & Hartse, 2020; Gasparyan et al., 2015). Even though peer review is advertised in most spam messages, which assure that the article will be subjected to that process (Excerpt 3.5), there are other signs (like rapid publication acceptance and prompt publication) which warn against the rigour of such peer review and cast doubt on whether proper peer review has actually been conducted. A short peer-review time is indicative of journal poor quality (Cobey et al., 2018; Gasparyan et al., 2015; Memon, 2019). When writing a research paper, scholars take for granted that they will be involved in a submission–acceptance/rejection–revision/rejection–publication cycle that may take months. It may also be the case that, after a long time of back-and-forth revisions, the article is eventually rejected. As discussed in the literature (e.g. Beall, 2016; Kurt, 2018), this does not happen with sub-par journals because it is contrary to the business model of predatory publishers, who are more interested in generating as much revenue as possible.

A last and relevant category is "bibliometrics" which, broadly speaking, entail bogus metrics and other misleading data (Excerpts 3.6a–e). Spam messages forwarded by predatory publishers claim high impact factors, which are usually self-calculated (Soler & Cooper, 2019), fake (Bloch, 2021), neither legitimate (Cobey et al., 2018) nor derived from recognised sources (Byard, 2016). They may also state dubious impact factors, or factors from unknown indexing agencies (Excerpt 3.6c). Very often, impressive (if not, false) figures are given; in most cases, these are impossible to verify. The "indexing" of predatory journals on databases such as Google Scholar or DOAJ, or academic social networks such as ResearchGate (see Excerpt 3.6d), nurtures "a discourse of authenticity" (Pecorari, 2021, p. 352), particularly among inexperienced and junior researchers, who are not yet familiar with those indexes that are reputable in their areas and who are not yet aware of the fact that these databases and platforms incorporate almost any (published or draft) study that is available on line—hereby diminishing their use for bibliometric analysis (Gasparyan et al., 2015). Lastly, other indicators like references to digital object identifiers (DOIs) or copyrights (see Excerpts 3.6a and 3.6e) are misleading to authors because they might be taken as evidence for journal quality—that is, DOIs are issued for article identification, and by accepting open access publication, authors, as copyright holders, consent to free distribution via open-content licences such as Creative Commons.

5 What Are the Implications for Open Access and the Dissemination of Knowledge?

From the extant literature and the findings of this study, it can be claimed that predatory journals appear to clearly contribute to the dissemination of non-validated knowledge in academia; therefore, causing damage to true science, researchers, and society. The implications of predatory publishing for the open access movement and the dissemination of scientific knowledge are multifarious and, at the same time, interrelated. They may be grouped into three main realities as discussed in the next subsections.

5.1 Spread of Non-validated Knowledge

Open access and questionable (even lack of) peer review contribute to strengthening the spread of "junk science" (Forero et al., 2018, p. 586) and the validation of "bogus experts" (Byard, 2016, p. 392). As I argued elsewhere (Bocanegra-Valle, 2017), particular research areas (like Applied Linguistics, or the Humanities in general) are more likely to be impacted by predatory practices if compared with journals in the hard or natural sciences. One reason for this is that there are many national and international reputed journals that offer free and unrestricted on line access to full-text articles. Another reason is that these other areas have a longer tradition in the dissemination of research findings with well-established quality benchmarks.

As predatory practices stand, anything can be published for a fee, "even the most bizarre theories with inadequate or no scientific validation" (Byard, 2016, p. 392). This means that non-validated knowledge can be published easily and disseminated rapidly—thereby "polluting" (Beall, 2016; Gasparyan et al., 2015; Shamseer et al., 2017) scholarly literature, and undermining the scientific conversation (Elmore & Weston, 2020). The consequences of this practice may be disastrous for scientific knowledge because if fringe theories and invalid sources are trusted by unwary authors, misinformation may spread out of control, spoof articles may be legitimised, and fraudulent research may be given a safe conduct to freely transit academia and society—by way of example, it is not difficult to imagine the impact of inaccurate health-related data if disseminated on line without control. In this regard, Pflugfelder (2022, p. 309) notes that predatory journals function as "evidence engines" that "thrive in a state of posttruth discourse" because (i) actors connected to larger conspiracy theories can use fraudulent articles to circulate their erroneous (usually ideologically driven) beliefs; and (ii) the public use of these articles will make illegitimate claims become evidence—therefore helping to promote pseudoscience.

On a final note, confidence in research literature may be undermined because authors may misunderstand the distinction between predatory and legitimate journals, thereby questioning the scientific process and losing public faith in research literature (Ferris & Winker, 2017).

5.2 Misconduct and Misdirection

It has been claimed that predatory journals are "veritable reservoirs of author misconduct" (Beall, 2013, p. 48; see also Byard, 2016). Besides absent or questionable peer review procedures and ethical standards (issues that have been discussed in this chapter), practices of research misconduct include all types of plagiarism (direct, patchwork, translational, or self-plagiarism), data and image manipulation, or retraction policy. Because articles are not subjected to a rigorous peer review and knowledge is not validated, misconduct is another threat to scholarly publication. It casts further doubts on the study that has been published and thus leads readers to approach findings with caution and to avoid the citation of research that may appear suspicious.

Predatory practices also involve a "strategy of misdirection" (Pecorari, 2021, p. 356) or "misrepresentation" (Ferris & Winker, 2017, p. 280); that is, scholars are driven into "a sort of fallacious thinking" (Pecorari, 2021, p. 356) because fraudulent journals contain signs that intentionally create a misleading impression and that scholars recognise from legitimate journals (like some of the categories in the analysis of this study—e.g. the use of invalid self-reported metrics). Misrepresentation has been identified by Ferris and Winker (2017) as a main ethical issue in academic publishing because, firstly, these journals misrepresent who they are and the services they offer; and, secondly, authors misrepresent their research by choosing to publish in predatory journals and listing them on their CVs as if they had been published in legitimate journals.

5.3 Impact on Researchers and Global Scholarship

To assume that deceptive journals prey on vulnerable scholars, or that researchers publish their work in these journals because they are naive or unconcerned, is a simplistic way to address the problem. The institutional and contextual factors that determine scholars' publication decisions are diverse and complex (Mills & Inouye, 2020). These include, among others, publication incentives and pressures on individuals, the relatively underdeveloped nature of a scholarly research publication culture, the role played by these journals in providing effective outlets for research, or additional hurdles like English language proficiency (Bloch, 2021; Collyer, 2016; Memon, 2019; Mills & Inouye, 2020). Xia et al. (2015) found that there was a preponderance of young and inexperienced researchers from developing countries, especially from India, Nigeria, and some African and Middle East countries, who were in serious need of becoming visible in academic circles and increasing their CVs in a very short time by building their list of publications quickly and easily. These findings concur with Forero et al. (2018) and Kurt (2018), for whom inexperienced young scholars from developing countries with a need to advance their careers are the most vulnerable victims of these practices.

It may also be the case that when some countries systematically publish their research in fake journals, articles will decrease in top-tier journals, therefore lowering their position in global academic ranking systems (Forero et al., 2018). In the same vein, Kurt (2018) warns of research published in predatory journals becoming "lost science" (p. 143)—that is, published research may not be located via formal academic searches and this will inevitably lead to the decrease of the research profile of the researcher's institution and country as well as the loss of future funding.

Likewise, Ferris and Winker (2017) raise a relevant ethical concern: the illegitimate consideration of a journal will question the legitimacy of the research published therein; thus, wasting funding and resources.

To my understanding, all these circumstances create a divide between mainstream authors, who publish in legitimate journals and can therefore be trusted, and authors, usually from emerging or developing countries, who publish untested findings in dubious or predatory journals; the former contribute to the expansion of validated knowledge, whereas the latter promote invalid or fraudulent science. These publication patterns might help to create and sustain the inequalities of global academic knowledge production between mainstream and non-mainstream countries, so that scholarship from "emerging markets of scientific research" (Shamseer et al., 2017, p. 2) may be systematically alienated or undervalued.

6 Conclusions

Different voices have claimed the need to promote scholarly publishing literacy or educate authors in science writing in order to improve their understanding of predatory practices and recognise publishing fraud (e.g. Beall, 2012; Memon, 2019; Soler & Wang, 2019). Aligning with this literature, this study has sought to identify the main features that characterise predatory publishing to generate interest and convince authors to submit their work. It has also assessed deceptive editorial practices that may flout mainstream academic publication standards and put the dissemination of validated knowledge and open access at stake. Specifically, the analysis of predatory emails has shown that predatory discourse overuses boastful language and relies on self-promotional explicit references to high quality in order to persuade authors and achieve publishers' predatory aims. The implications of deceptive journals and predatory practices for knowledge dissemination and the open access movement have been discussed with particular reference to the spread of suspicious knowledge, misconduct practices and misdirection strategies, and impact on scholars (usually inexperienced and from emerging countries) who need to build up their publication portfolios and gain visibility in global academia.

The main recommendations that derive from this study are three: avoid distrusting open access journals from the outset ("predatory" and "open access" are not interchangeable terms); disbelieve prompt and swift publication; and make sure that peer review is safeguarded. Also, lessons learned from this study point at the need to examine journal websites, particularly for unrealistic author-pay processing or publication; keep a wary eye on misleading data (like apparently well-known index lists and databases); investigate who has previously published in that journal and if the journal is often cited in a particular field; and look for reputed scholars who are relevant in that field, either as authors or members of scientific/editorial boards.

Doubtlessly, all researchers and academics will benefit if they gain the skills and awareness to differentiate legitimate from scam journals. If they are able to identify the "red flags" of deceptive publishing practices, they will also be ready to identify high-quality venues and make informed choices of the most appropriate journal for their research. Becoming alert of the existence of predatory publishers and learning the signs of fraudulent practices offer bottom-up benefits in academia. Firstly, it will prevent early-career researchers from putting faith in a piece of research whose publication will not compensate for the effort in terms of curriculum reward. It will also help experienced researchers in their role of supervisors and mentors to streamline junior researchers' work and results. At the institutional level, evaluating and tenure committees in charge of examining scholars' credentials will be able to conduct informed assessments on the quality of the published research, and based on journal legitimacy.

This study is a first exploration into the ways academic research is impacted by predatory practices, although further investigation is needed to have a better understanding of the academic dynamics in which predatory publishing is flourishing. Further explorations of how authors are engaged in predatory practices could help us better understand how engagement strategies are used to create a persuasive appeal, above all on junior researchers (Bocanegra-Valle, 2023). Likewise, the multimodal features and paratextual elements that prevail in predatory messages are issues that could be addressed. For these purposes the group of messages under analysis in this chapter could be extended. Conducting ethnographically-oriented analyses is another direction for further research. Junior and experienced researchers as well as journal editors and publishers could be approached to tap their views and concerns on predatory publishing. Also, scholars who have actually published in predatory journals could be surveyed and interviewed to gather more insights into the reasons and expectations when submitting their work for eventual publication.

Acknowledgements This study is a contribution to the project *Digital Genres and Open Science*, funded by the Spanish Ministry of Science and Innovation (PID2019-105655RB-I00, funded by MCIN/AEI/10.13039/501100011033)

References

Beall, J. (2012). Predatory publishers are corrupting open access. *Nature, 489*(7415), 179.

Beall, J. (2013). Medical publishing triage—Chronicling predatory open access publishers. *Annals of Medicine and Surgery, 2*(2), 47–49.

Beall, J. (2015). *Criteria for determining predatory open-access publishers.* https://beallslist.net/wp-content/uploads/2019/12/criteria-2015.pdf

Beall, J. (2016). Dangerous predatory publishers threaten medical research. *Journal of Korean Medical Science, 31*, 1511–1513. https://doi.org/10.3346/jkms.2016.31.10.1511

Beall, J. (2017). *Beall's list of predatory open access publishers 2017.* https://scholarlyoa.com/2017/01/03/bealls-list-of-predatory-publishers-2017/

Bloch, J. (2021). *Creating digital literacy spaces for multilingual writers.* Multilingual Matters. https://doi.org/10.21832/BLOCH0794

Bocanegra-Valle, A. (2017). How credible are open access emerging journals? A situational analysis in the humanities. In M. Cargill & S. Burgess (Eds.), *Publishing research in English as an additional language: Practices, pathways and potentials* (pp. 121–149). University of Adelaide Press.

Bocanegra-Valle, A. (2019). Building a reputation in global scientific communication: A SWOT analysis of Spanish humanities journals. *Canadian Journal of Sociology, 44*(1), 39–66. https://doi.org/10.29173/cjs28935

Bocanegra-Valle, A. (2020). Researching academic genres, language and discourse with computer-assisted qualitative data analysis software. In M. L. Carrió-Pastor (Ed.), *Corpus analysis in different genres: Academic dis-*

course and learner corpora (pp. 146–166). Routledge. https://doi.org/10.4324/9780367815905

Bocanegra-Valle, A. (2023). Engaging in predatory practices: How editors persuade prospective authors. *Círculo de Lingüística Aplicada a la Comunicación, 93*, 117–129. https://doi.org/10.5209/clac.82441

Byard, R. W. (2016). The forensic implications of predatory publishing. *Forensic Science, Medicine, and Pathology, 12*, 391–393. https://doi.org/10.1007/s12024-016-9771-3

Cobey, K. D., Lalu, M. M., Skidmore, B., Ahmadzai, N., Grudniewicz, A., & Moher, D. (2018). What is a predatory journal? A scoping review. *F1000Research, 7*, 1001. https://doi.org/10.12688/f1000research.15256.2

Collyer, F. M. (2016). Global patterns in the publishing of academic knowledge: Global north, global south. *Current Sociology, 66*(1), 56–73. https://doi.org/10.1177/0011392116680020

Elmore, S. A., & Weston, E. H. (2020). Predatory journals: What they are and how to avoid them. *Toxicologic Pathology, 48*(4), 607–610. https://doi.org/10.1177/0192623320920209

Fazel, I., & Hartse, J. H. (2018). Reconsidering 'predatory' open access journals in an age of globalised English-language academic publishing. In M. J. Curry & T. Lillis (Eds.), *Global academic publishing. Policies, perspectives and pedagogies* (pp. 200–213). Multilingual Matters. https://doi.org/10.21832/CURRY9238

Fazel, I., & Hartse, J. H. (2020). Gray areas of academic publishing. 'Predatory journals' under the lens. *Journal of English for Research Publication Purposes, 1*(2), 184–194. https://doi.org/10.1075/jerpp.20008.faz

Ferris, L. E., & Winker, M. A. (2017). Ethical issues in publishing in predatory journals. *Biochemia Medica, 27*(2), 279–284. https://doi.org/10.11613/BM.2017.030

Forero, D. A., Oermann, M. H., Manca, A., Deriu, F., Mendieta-Zerón, H., Dadkhah, M., Deshpande, S. N., Wang, W., & Cifuentes, M. P. (2018). Negative effects of "predatory" journals on global health research. *Annals of Global Health, 84*(4), 584–589. https://doi.org/10.29024/aogh.2389

Gasparyan, A. Y., Yessirkepov, M., Diyanova, S. N., & Kitas, G. D. (2015). Publishing ethics and predatory practices: A dilemma for all stakeholders of science communication. *Journal of Korean Medical Science, 30*, 1010–1016. https://doi.org/10.3346/jkms.2015.30.8.1010

Kurt, S. (2018). Why do authors publish in predatory journals? *Learned Publishing, 31*, 141–147. https://doi.org/10.1002/leap.1150

Laine, C., & Winker, M. A. (2017). Identifying predatory or pseudo-journals. *Biochemia Medica, 27*(2), 285–291. https://doi.org/10.11613/BM.2017.031

Memon, A. R. (2019). Revisiting the term predatory open access publishing. *Journal of Korean Medical Science, 34*(13), e99. https://doi.org/10.3346/jkms.2019.34.e99

Mills, D., & Inouye, K. (2020). Problematizing 'predatory publishing': A systematic review of factors shaping publishing motives, decisions, and experiences. *Learned Publishing, 34*(2), 89–104. https://doi.org/10.1002/leap.1325

Pecorari, D. (2021). Predatory conferences: What are the signs? *Journal of Academic Ethics, 19*, 343–361. https://doi.org/10.1007/s10805-021-09406-4

Pflugfelder, E. H. (2022). Evidence engines: Common rhetorical features of fraudulent academic articles. *Written Communication, 39*(2), 303–331. https://doi.org/10.1177/07410883211069332

Shamseer, L., Moher, D., Maduekwe, O., Turner, L., Barbour, V., Burch, R., Clark, J., Galipeau, J., Roberts, J., & Shea, B. J. (2017). Potential predatory and legitimate biomedical journals: Can you tell the difference? A cross-sectional comparison. *BMC Medicine, 15*(28), 1–14. https://doi.org/10.1186/s12916-017-0785-9

Soler, J., & Cooper, A. (2019). Unexpected emails to submit your work: Spam or legitimate offers? The implications for novice English L2 writers. *Publications, 7*(1), 6–11. https://doi.org/10.3390/publications7010007

Soler, J., & Wang, Y. (2019). Linguistic differences between well-established and predatory journals: A keyword analysis of two journals in political science. *Learned Publishing, 32*(3), 259–269. https://doi.org/10.1002/leap.1244

Xia, J., Harmon, J. L., Connolly, K. G., Donnelly, R. M., Anderson, M. R., & Howard, H. A. (2015). Who publishes in "predatory" journals? *Journal of the Association for Information Science and Technology, 66*(7), 1406–1417. https://doi.org/10.1002/asi.23265

7

Between Infotainment and Citizen Science: Degrees of Intended Non-expert Participation Through Knowledge Communication

Jan Engberg

1 Introduction

Digital genres used for making research projects and research results visible to wide and diverse audiences are the object of much present-day research like, for instance, the InterGedi project, from which the present book originates (http://intergedi.unizar.es/). Choosing this type of communication as object of study, the multifunctionality of digital texts in science communication and the addressing of multiple and different stakeholders comes into focus. It becomes relevant to go beyond the mere description of the textual patterns underlying the respective digital genres that we need insights into in order to educate and guide scientists that want to enter into this type of communication. The focus is widened towards seeing the publication practice represented in different genres as embedded in a varied set of overall goals of communicatively creating

J. Engberg (✉)
Aarhus Universitet, Aarhus, Denmark
e-mail: je@cc.au.dk

R. Plo-Alastrué, I. Corona (eds.), *Digital Scientific Communication*,
https://doi.org/10.1007/978-3-031-38207-9_7

relations between the researcher and the surrounding public. With this larger scope, we may achieve insights into the background of the genres and may thus better understand the characteristics of the genres and their underlying textual patterns.

A good example of this wider scope is found in the concluding part of Rosa Lorés' analysis of values in the Homepages and About pages of some Horizon 2020 research projects with a focus upon the evaluative language (Lorés, 2020, p. 8):

> Finally, there seems to be a link between the use of positive evaluative language and *the promotion of the accessibility of science*: to make science open is not simply a question of making it accessible to the citizen but also a question of *promoting interest in science*. And one way of doing so is by evaluating research projects and their outcomes as positive, as having an impact in society. In all, the study of H2020 websites as a type of digital scientific writing practice has contributed to our understanding of how scientific discourse is taking shape *in a context of civil science*, where policies are increasingly undertaken to *make science accessible to the public*, whether expert or not (emphasis added).

The interesting aspect of the citation from the point of view of my contribution here is that a number of potential goals and contexts of the investigated digital scientific writing practice concerning the intended relation between science and the general public are extracted from the results of the analysis: *to make science open*, *to make it accessible*, *promoting interest in science*, *context of civil science*. A general common trait in the goals is that they are connected to the asymmetric relation concerning knowledge between science and the surrounding society as addressees of the digital writing practice, which should be handled in a way that opens science up to society and its citizens. As the cited article and many other contributions from the InterGedi project show, it has proven very fruitful to use this lens when analysing the mentioned types of websites in the project. In my contribution, I want to depart from these insights and proceed further down this alley, albeit in a chiefly conceptual way and with more focus upon the context of the publication practice than on the generic patterns. Starting from the policy-level concept of 'open science' as a societal goal, I propose a continuum for categorising scientific writing efforts, in which we may distinguish different types of science communication according to their degree of participatory ambition. I want to propose criteria for distinguishing between scientific writing efforts according to the intended relation between science and the general public reaching from (mere) entertainment to full inclusion as participating researchers. The idea is to describe some central points on this continuum and their internal relations as prototypes in order to offer a basis for empirical linguistic and multimodal analyses of scientific writing practices. On this basis, a central aspect of the contextual backdrop upon which the different genres in this field exist and function may be made available for analysis.

Concerning the communicative efforts I have in mind, when I propose the contextual description, I do not limit myself to the project presentation of digital genres investigated by the InterGedi project, although the deliberations should also be relevant for this type of texts. Instead, I want to propose a description of contextual factors with relevance for all types of science communication aimed at going beyond communicating with the experts' own peer group.

The structure of the chapter is first to introduce in Sect. 2 the basic characteristics of the knowledge communication approach that I use when investigating instances of communication characterised by knowledge asymmetries. The main part of the chapter (Sect. 3) will discuss different concepts forming a continuum of communicative efforts aiming at different degrees of participation in science by the receivers, distinguishing between policy concepts (Sect. 3.1) and concepts of knowledge communication (Sect. 3.2). Finally, in the concluding Sect. 4, I sum up the suggested continuum and put it into context.

2 The Knowledge Communication Approach as Framework

The knowledge communication approach has been developed since 2004 in collaborative work among a group of researchers at the Aarhus University.[1] The basic idea was to develop the traditional concepts of

[1] Cf. Porup Thomasen (2015), Kastberg (2019) and Engberg et al. (2023), for details on the approach.

Language for Specific Purposes (LSP) and specialised communication with their main focus upon language and (situated) texts by shifting the attention to (expert) knowledge as the central characteristic of this type of communication. The following definition sketches the main characteristics of the approach as I am going to use it in this chapter:

> The study of Knowledge Communication aims at investigating the intentional and decision-based communication of *specialized knowledge* in professional settings (among experts as well as between experts and non-experts) with a focus upon the interplay between knowledge and expertise of *individuals*, on the one hand, and knowledge as a *social phenomenon*, on the other, as well as the coping with *knowledge asymmetries*, i.e., the communicative consequences of differences between individual knowledge in depth as well as breadth. (Engberg, 2016, p. 37; emphasis added)

The aim of this framework is to investigate the intentional communication of specialised knowledge in professional settings performed as part of the practice of being an expert. The aspect of knowledge as being simultaneously an individual and social phenomenon is not central to the deliberations in this chapter and will therefore not be dealt with here.[2] Instead, I want to focus upon the aspect of knowledge asymmetry. Participants in knowledge communication are either all experts or a mixture of experts and non-experts. There is no categorial difference between expert–expert communication and expert–non-expert communication. In both cases, meaning in communication emerges from the individual knowledge bases of the participants in a constructivist process, in which the receivers make sense of the communicative input on the basis of their previous knowledge and their perception and consideration of the communicative context (Engberg, 2022, pp. 68–73). The difference between the two types of knowledge communication thus lies in the breadth, depth and concrete organisation of the individual knowledge bases, which naturally have an impact upon the characteristics of the communication: The more the knowledge bases of the participants in the communication are similar, the more communicators may rely upon

[2] Cf. Engberg (2015), for discussions of the analytical repercussions of this characteristic.

their existing knowledge base to understand communicative input; on the other hand, the more differences exist between the involved knowledge bases, the more new constructions have to be performed in the course of understanding the communicative input. This aspect is what is meant by the concept of knowledge asymmetry. The concept is central for this chapter as it is related, as highlighted above, to how to make science open, make it accessible and promote interest in science in a context of civil science (Lorés, 2020). These ideas rely on the fact that differences exist between individual knowledge in depth as well as breadth and organisation and that (expert) communicators want to cope with this situation, that is, the second aspect in the definition above.

In the knowledge communication approach, we depart from the idea that knowledge is always individual (as well as a social phenomenon, cf. the first aspect in the definition above). As the individual knowledge is constructed from experiences and learnt insights, the individual knowledge bases will to some extent always be asymmetric and actually function as "a potent motivational device" for entering into knowledge communication (Kastberg, 2019, p. 99). For our interest in expert communication, to which scientific writing practices belong, however, not the fact of asymmetry as such is important. Instead, from this point of view, the fact that such asymmetries are made communicatively visible is actually the analytically interesting aspect:

> Knowledge asymmetry is a relation which is produced in communication. The discursive construction of knowledge asymmetry is observable via perturbations. Knowledge asymmetry becomes communicatively salient where a single-plane distinction is observed between the knowledges of 'alter' and 'ego'. The distinction is one which allows one or more positions to appreciate the knowledge of 'alter' and 'ego' as being non-identical under the same sortal. (Kastberg, 2011, p. 145)[3]

The normal situation in recognised expert–lay communication is that the expert has realised the difference and makes it visible in communication, for instance through explanations or indications of the expert role of the

[3] Kastberg (2019, pp. 86–89) explains this as inspired by a geometrical view of asymmetry.

sender. Thus, the knowledge asymmetry is discursively constructed by the sender of the communicative effort (Kastberg, 2019, p. 93). The sender presents a perspective of the concepts in the conversation as central and as one where the sender has knowledge relevantly different from that of the addressees in depth, breadth or organisation. The aspect of knowledge asymmetry that interests me in the deliberations in Sect. 3 is the degree to which an expert sender in the discursive construction presents the difference/asymmetry concerning depth, breadth and/or organisation as something that may be overcome through mutual participation among communicators, that is, to what degree non-expert participants are invited to get closer to constructing expert-like knowledge structures. For this purpose, I will in the following section present the continuum of knowledge communication types mentioned above in Sect. 1 as a way of distinguishing communicative efforts along these lines.

3 Degrees of Participation—Differentiating Knowledge Communication Types

The concepts chosen for this chapter are those that emerge consistently in the discussions about science having to be openly accessible in democratic societies and that are also treated widely in the works from the InterGedi project. In these discussions, I see two distinct classes of concepts involved: Those connected to research policy in general and those connected to the communication of scientific results across knowledge asymmetries, that is, concepts at the level of knowledge communication. As the focus of this chapter is upon sketching the backdrop for the genres of scientific writing practices, the set of concepts connected to knowledge communication are the most central ones here. However, I will briefly introduce the first class of concepts in order to make the distinction and its influence upon the knowledge communication concepts clear. In this context, I will consider the concepts of 'open science' and 'citizen science' and their differences and similarities in Sect. 3.1. In the main part of these deliberations (Sect. 3.2), I will look at the knowledge communication concepts of 'infotainment', 'dissemination', 'popularisation', 'pedagogic communication' and (again here) 'citizen science'.

3.1 Concepts at Policy Level—'Open Science' and 'Citizen Science'

The concept of 'open science' is defined, for instance, by the OECD:

> Open science encompasses unhindered access to scientific articles, access to data from public research, and collaborative research enabled by ICT tools and incentives. Broadening access to scientific publications and data is at the heart of open science, so that research outputs are in the hands of as many as possible, and potential benefits are spread as widely as possible:
>
> - Open science promotes a more accurate verification of scientific results. By combining the tools of science and information technologies, scientific enquiry and discovery can be sped up for the benefit of society.
> - Open science reduces duplication in collecting, creating, transferring and re-using scientific material.
> - Open science increases productivity in an era of tight budgets.
> - Open science results in great innovation potential and increased consumer choice from public research.
> - Open science promotes citizens' trust in science. Greater citizen engagement leads to active participation in scientific experiments and data collection.
>
> https://www.oecd.org/sti/inno/open-science.htm

As is visible from the citation from the OECD website, the concept of 'open science' is closely linked to a specifically 'open' way of carrying out research: Knowledge generated from (especially publicly funded) research is not a commodity belonging to the researchers, but a public good that should be shared, both due to its value for society (often subsumed under the headline of 'democratisation of science' [Kastberg, 2019, p. 81]) and in order to enhance the quality of the research itself (Luzón & Pérez-Llantada, 2019, p. 3). Looking at the bullet points, we find both of these aspects. The first four bullet points are about enhancing scientific performance in a society, whereas the fifth one centrally talks about the value of openness for the citizens of the society. So based on this presentation of the concept, 'open science' for the OECD has its focus upon enhancing the efficiency of science (*accurate verification, no reduplication, increased productivity, increased innovation potential*), but does also include the value of openness for the society at large (*trust in science, active participation*).

In my view there is a logical break between the first four bullet points and the last one. To put it bluntly, the fact that more researchers talk to each other and share the collection and use of scientific material does not lead to citizens trusting science. This means that the open science concept contains two aspects: On the one hand, it treats the general policy decisions for the organisation of research and, on the other hand, it is oriented towards the actual democratisation of science, that is, making scientific research processes and products accessible to the wider public. These conceptual characteristics are represented in Fig. 7.1.

The other concept to be treated here is the concept of 'citizen science'. We depart from the EU's characterisation of the concept on a dedicated website:

> What is citizen science?
> - Citizen science is any activity that involves the public in scientific research and thus has the potential to bring together science, policy makers, and society as a whole in an impactful way.
> - Through citizen science, all people can participate in many stages of the scientific process, from the design of the research question to data collection and volunteer mapping, data interpretation and analysis, and to publication and dissemination of results.
> - Citizen science is also an approach of scientific work that may be used as a part of a broader scientific activity.
>
> https://eu-citizen.science/

Fig. 7.1 Policy concept of 'open science' related to participation in science and its components

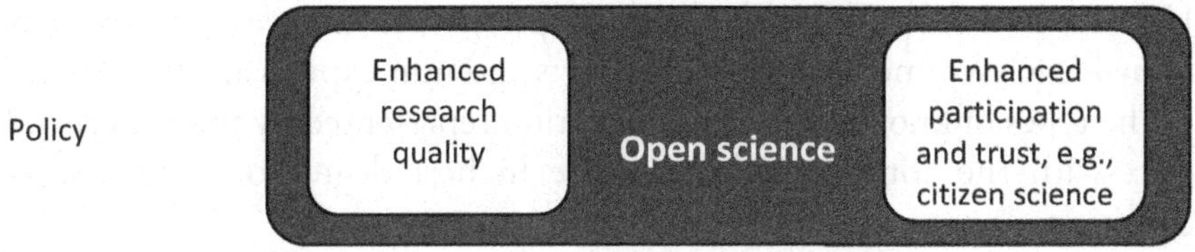

Fig. 7.2 Policy concepts of 'open science' and 'citizen science' related to participation in science

The key aspects in this characterisation of the concept from the point of view of this chapter are that citizen science *involves the public*, *brings science and the public together* and *inspires public participation*. All of these goals have to do with the part of the 'open science' concept related to enhance participation. I see the concept 'citizen science' not only, but also as a policy concept as it is about how science should be organised to open up towards society. Thus, it can be seen as fleshing out what is said about 'open science' in the fifth bullet point of OECD's characterisation above as indicated in the addition in Fig. 7.2, compared to Fig. 7.1. As I propose to see 'citizen science' also as a communication-oriented concept, I will go deeper into the concept in Sect. 3.2.

The central policy-level concept is thus 'open science'. It is the overarching concept to designate a societal push towards two meanings of openness of science, but the characteristics of the concept influence the different genres in the writing practice only in a rather indirect way. The concept highlights the fact that science in its design and communication with society has multiple addressees, as its two aspects do not address the same goals or parts of society. Hence, the ensuing scientific writing practice must be characterised by the duality of opening science up to fellow scientists (for enhancing research quality) and simultaneously opening science up to the society at large (for enhancing participation and trust).

3.2 Concepts at the Level of Knowledge Communication

In this section, I present different knowledge communication concepts according to the degree of participatory ambition indicated in them.

Hence, the idea is to rank the concepts in accordance with the way knowledge asymmetries between experts and non-experts are constructed in the types of knowledge communication represented by the concepts. I start with the concept indicating the highest degree of participatory ambition.

Citizen Science

As indicated above, 'citizen science' is about involving the public in scientific work. Thus, from the point of view of participation, the participatory ambition is very high here. Apart from seeing it as a concept at policy level concerning the organisation of scientific work, I propose to also see it as a knowledge communication concept. For when a research project opts for a citizen science setup, it communicates through choosing this high level of participatory ambition that the experts in the project want to construct the asymmetric knowledge-oriented relation to non-experts at an equal footing, inviting the non-experts to approach the knowledge systems of the experts in a way that enables experts and non-experts to cooperate in research. The scientific writing practice under this concept must reflect this level of ambition.

The characteristics mentioned above are visible in ten principles for citizen science developed by "an international community of citizen science practitioners and researchers to set out their shared view of the characteristics that underpin high-quality citizen science" (Robinson et al., 2018, p. 27):

1. Citizen science projects *actively involve citizens* in scientific endeavour that generates new knowledge or understanding.
2. Citizen science projects have a *genuine science outcome*.
3. Both the professional scientists and the *citizen scientists* benefit from taking part.
4. Citizen scientists may, if they wish, *participate* in multiple stages of the scientific process.
5. Citizen scientists *receive feedback* from the project.

6. Citizen science is considered *a research approach like any other*, with limitations and biases that should be considered and controlled for.
7. Citizen science project data and metadata are made publicly available and where possible, results are published in an open-access format.
8. Citizen scientists *are acknowledged* in project results and publications.
9. Citizen science programmes are evaluated for their scientific output, data quality, *participant experience* and wider societal or policy impact.
10. The leaders of citizen science projects take into consideration legal and ethical issues surrounding copyright, intellectual property, data-sharing agreements, confidentiality, attribution and the environmental impact of any activities. (Robinson et al., 2018, pp. 29–30; emphasis added)

I have highlighted some of the most central formulations in the principles: The high level of participatory ambition is signalled by the principle to *actively involve citizens*, which is exemplified through formulations like *participate*, *receive feedback* and *are acknowledged*, as well as by talking about the non-expert participants as *citizen scientists*. Furthermore, the intended equal footing is indicated by the requirements that citizen science projects have a *genuine scientific outcome* and is *a research approach like any other*, and that the *participant experience* is evaluated. The basic idea of citizen science is thus to treat the knowledge asymmetry as a difference in knowledge that may be overcome—to the extent that non-experts are invited and included as full members of the research team, albeit with other tasks than the experts.

As indicated above, I see 'citizen science' also as a knowledge communication concept. A central reason is that realising the principles cited above means that it will be necessary to level out existing knowledge asymmetries. On a continuum showing the degree of participation in research intended, I would place 'citizen science' at the top end (cf. Fig. 7.3 at the end of Sect. 3). The focus here is on creating responsibilities of the citizen scientists that in my view will also result in creating an emotional bond between experts and non-experts and thus to react to the requirement of the policy concept of 'open science' to enhance participation and trust from society at large.

Popularisation, Pedagogic Communication and Dissemination

In the field of research into scientific writing practice across knowledge asymmetries, there is a tendency to use words like *popularisation*, *dissemination*, *democratisation*, etc., as synonyms. However, I think that this approach is less than optimal, because we lose the possibility of making distinctions in the description of the context of the scientific writing practices that may help us understand better differences between genres and differences between the requirements in different communicative constellations. I have therefore proposed to make a distinction between the concepts of 'popularisation', 'pedagogic communication' and 'dissemination' (Engberg, 2020a, 2020b; Engberg et al., 2018; Engberg & Heller, 2020).

Point of departure was the characterisation of popularisation by Maurizio Gotti:

> There is a basic consensus, however, as to the role of this process, which is usually identified with *the conveyance of specialist knowledge for information purposes.* […] The main criterion for distinguishing between fully specialized texts and popularisations is the *different audience targeted.* Popularization in fact addresses not an expert group within the discipline but an audience of non-specialists. […] Publications written primarily for non-specialists operate at no fewer than two different levels: *pedagogic texts* and *popularizations.* (Gotti, 2014, p. 16)

The characteristic distinguishing what Gotti calls popularisation texts from fully specialised texts is a new context, chiefly defined by the change of addressees leading to a more pointed knowledge asymmetry (recontextualisation; Calsamiglia & Van Dijk, 2004; Liebert, 2002). Change of context also distinguishes 'popularisation' from 'citizen science', as we have seen that the latter intends to stay inside the research context and educate the non-experts to be able to function as citizen scientists, whereas the context of popularisation is that of informing non-participating addressees. However, this difference is one of degree rather than of category. Furthermore, I follow Gotti's distinction between pedagogic texts and popularisations, again based on contextual characteristics: Both concepts designate knowledge communication efforts across knowledge asymmetries. But the contextual difference lies in the fact that pedagogic communication is connected to a specific instructional institutional context governing the relation between the participants (*Lehr-Lern-Diskurs* Redder, 2002). The aspects of the instructional institutional context consist, for instance, of specific institutional goals relevant for specific professions to be tested through exams, which popularisations lack. Popularisations, on the other hand, appear in situations where the expert sender cannot rely upon an instructional institutional role but must generate interest among addressees from instance to instance, for example, when institutions like the European Parliament inform EU citizens about their role and the legal system underlying their activity.

In both concepts, providing information is seen as the main purpose of the communicative effort. However, when we supplement this characteristic with the dimension of participatory ambition, we can insert a further distinction of what Gotti terms 'popularisation' with relevance for the intended relation across the knowledge asymmetry. For this purpose, I want to refer to a basic distinction between different types of action suggested by Jürgen Habermas. In his philosophy, he is interested in how to uphold a democratic society based on communicative interaction leading to mutual understanding. He distinguishes between two different kinds of action: *Strategic action* and *communicative action* (Habermas, 1981, p. 385). In strategic action, the goal is to make the addressees do what the sender wants them to. The interaction is goal-oriented. Furthermore, and important for my argumentation here, strategic action implies a relation between interlocutors in which they both try to influence each other as opposed to the situation of communicative action, where the goal is to achieve an understanding comparable with the one between members of the same lifeworld (*Angehörige derselben Lebenswelt*, Habermas, 1981, pp. 386–387). In the terms of expert–non-expert knowledge communication, under the conditions of strategic action scientists talk to non-experts in order for the non-experts to do or understand something that is relevant for the scientists (and maybe also for the citizens, but that is not given in the concept). Non-experts stay non-experts, albeit somewhat wiser, but without achieving a closer relation to the scientists, without an intent of making the non-experts part of the scientists' lifeworld. In communicative action, on the other hand, the idea is that scientists aim at achieving understanding (*Verständigung*) among the non-experts, not only about what to do or what to basically understand, but also about underlying mechanisms and theories that will approach the non-experts to the lifeworld of the experts, and vice versa. In communicative action, the scientists also try to understand the non-experts, as well as possible, in order to build mutual understanding and reach agreement about worldviews.

Citizen science is a prime example of communicative action: Scientists and non-experts (here: citizen scientists) aim at reaching a stage through communication and other interaction where they—despite the knowledge asymmetry between them—can work together and take advantage of each other's insights and actually have the feeling of belonging to the same group of expertise, albeit at somewhat different levels. But even if the scientists do not want to go as far as actually working together with citizen scientists on mutual projects, they may want non-expert interlocutors to share so much common ground that the non-experts understand the reasons for what the scientists say and do. Compared to a more strategy-based interaction, where scientists merely tell non-experts about the consequences of their insights, the first-mentioned approach is closer to communicative action and the second approach is closer to strategic communication. Along these lines I propose to distinguish two different concepts of recontextualised, non-institutional information rendering: 'popularisation' tending more towards communicative action, on the one hand, and 'dissemination' tending more towards strategic action, on the other. An aspect concerning the intended relation between the participants distinguishing the two is that 'popularisation' due to its communicative characteristics of creating common ground often is characterised by traits of emotional bonding from the feeling of at least rudimentarily belonging to the same group: 'Popularisation' aims at not only telling the non-expert addressee about the expert knowledge, but also to make them like it by making it part of the non-experts popular culture (Engberg,

2020a, pp. 177–179). 'Dissemination' lacks this characteristic due to being closer to strategic action.

Concerning the continuum of participatory ambition, the distinction along the lines of communicative versus strategic action means that 'popularisation' is closer to the end of the continuum indicating a high degree of participation than 'dissemination'. I propose to position 'popularisation' and 'pedagogic communication' at the same level of participatory ambition. Both do not aim at turning non-experts into experts, but popularisation and didactic texts are at the same level of intended inclusion of the public in research, namely that of not actually including them, but at least giving them information so they can understand what is going on. The difference between the two lies in the fact that in pedagogic communication, the expert (= the teacher) typically has a focus on knowledge content, implying: "I want you to know what are the basic things in this field that we talk about, to build bridges into the field." Communication is carried out within the confines of a specific institution, implicating, for instance, that levels of expertise are tested through exams. 'Popularisation' lacks this institutional characteristic, although it shares the intent of building bridges into the field.

Infotainment

The last knowledge communication concept to be treated here is that of 'infotainment'. According to the *Oxford English Dictionary*, the word 'infotainment' means "broadcast material, which seeks to inform and entertain simultaneously; (also more generally) informative material presented in an entertaining way". Hence the central characteristic here is that of entertainment: Infotainment is entertainment, which is also informative. Information is oftentimes rather spin-off and is used very much in working with TV formats.

Concerning participation, it has the lowest degree of participatory ambition of the concepts presented here. The reason is that in many cases the construction of expert-like knowledge is not very central. As an example, TV documentaries from the field of history as analysed by Jaki (2018, 2020) may be mentioned. One of her central results is that in instances of knowledge communication of the infotainment type aspects of emotionalisation and entertainment play a more important role than the actual knowledge mediation. For instance, in history documentaries, re-enactments of historical actions typically play a central role, no matter whether the re-enacted historical action is actually central to the historical knowledge being conveyed at a specific point in the documentary. For instance, we may see the crowning of a person as king, whenever this person is mentioned, no matter whether the actual crowning is central to what is said at that moment. So, the use of the re-enactments is more a question of taking advantage of the fact that they generally are perceived as entertaining and therefore generate interest and make viewers stay tuned. Viewers may also learn something, but that is not the main point. In other words, as expert-like knowledge may be constructed by non-expert participants, albeit at a shallow level, there is some level of participatory ambition. But this ambition is considerably less prominent than what is the case in 'dissemination' as the neighbouring concept: In dissemination, the experts specifically want the non-experts to get to know something. In infotainment, this intended learning effect is less prominent. For this reason, I position the concept at the end of the continuum with the lowest degree of participatory ambition (cf. Fig. 7.3).

However, it is naturally the case that if you do good infotainment, with its elements of emotionalisation, then that might make non-expert addressees aware of specific insights that they may use for becoming interested in dissemination or popularisation efforts. Especially, this emotionalisation side of infotainment makes it relevant to, for instance, use infotainment as a tool that grabs the attention of people, and then lays the ground for following up with popularisation efforts in order to build bridges and emotional bonds. In this way, infotainment could be used in order to generate more interest in participation on the side of the non-experts.

With this, different knowledge communication concepts relevant for scientific writing practice in a broad sense have been distinguished and positioned on a continuum according to the participatory ambition

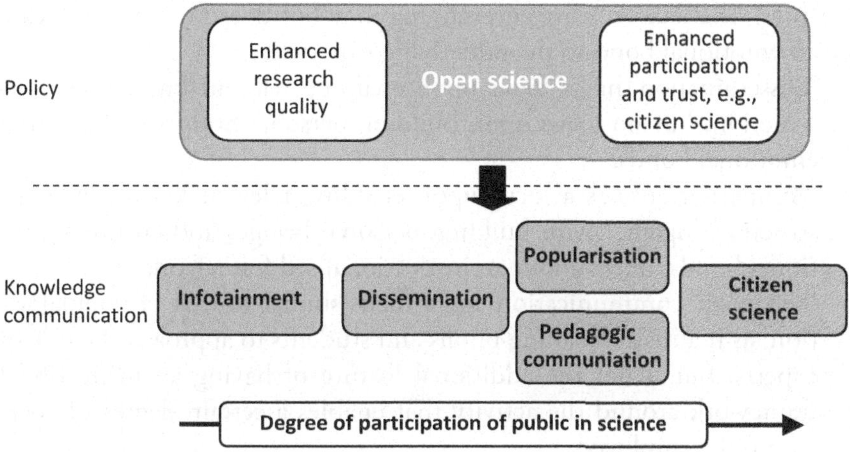

Fig. 7.3 Policy concepts and knowledge communication concepts related to participation in science

behind them. Furthermore, policy concepts have been distinguished from knowledge communication concepts and have been described as having a contextual role for the scientific writing practice. A next step (however, not to be taken here) could be to investigate different genres enclosed in projects like InterGedi in order to see which types of knowledge communication concepts in the contextual background they actually draw upon. In this context, the same project website could probably take advantage of types of knowledge communication from different points of the continuum depending on the exact function of a webpage or a section on a webpage. However, going deeper into this aspect would go beyond the confines and intentions of this contribution.

4 Concluding Remarks

Summing up the knowledge communication concepts presented in Sect. 3.2 and represented in Fig. 7.3, I propose the following short characteristics:

- 'Infotainment' has its focus on being entertaining and less upon enabling the addressees to construct more elaborate knowledge struc-

tures. The fact of being entertaining may but does not have to create an emotional bond to the underlying expertise.
- 'Dissemination' has a focus upon enabling relevant knowledge construction, but no focus upon building personal bridges and creating emotional bonds.
- 'Popularisation' has a focus upon enabling relevant knowledge construction together with building personal bridges and creating emotional bonds, but without an institutionalised framework.
- 'Pedagogic communication' has a focus similar to that of popularisation, as it aims at enabling pupils and students to approach the role of experts, but it has the additional feature of having an institutional framework around the activity that enables a certain degree of coercion to be employed.
- 'Citizen science' has a focus on full participation that enables the construction of expert-like knowledge structures and thus creates the strongest personal and emotional bonds between non-experts and experts.

From the point of view of the goal of the concept of 'open science' to enhance society's participation and hence trust in science, scientific writing practices carried out within citizen science is the most powerful tool. However, as it requires considerable effort and interest on the part of the non-experts, the other concepts presented above have important roles to play in connection with other communicative constellations. Although following Habermas' approach communicative action is more ethical than strategic action, in the reality of knowledge communication in society, for instance, the interests of the addressees may make them more prone to dissemination than to popularisation. As an example, in the Danish context of the COVID-19 pandemic, epidemiologists wanted to make the public aware of the kinds of discursive processes underlying the development of vaccines, in which insights are improved by being challenged in review processes. As these review processes were made public, thus demonstrating that disagreement among experts is normal, some non-experts drew the conclusion that if there was no final right answer to the research questions, on which everyone agreed, then science could not be trusted at all. In the terms above, I see this as an example of non-experts wanting to have a dissemination relation to the epidemiologists ('tell me what the sensible thing is to do, but not why and not how you reached your result'). The epidemiologists, on the other hand, wanted to build a popularisation relation across the knowledge asymmetry in order to enable the non-experts to make decisions on the basis of insights rather than mere instruction.

Finally, from the point of view of how to proceed with linguistic and discursive analyses based on the proposed continuum of knowledge communication concepts reaching from infotainment to citizen science, I see a few possibilities, as indicated above. Most centrally, the distinction between different concepts is based on the contextual feature of attributed degree of participatory ambition on the side of the sender. A next step could be to compare the textual practices connected to the different concepts in project websites in order to find out whether there are conventional linguistic indicators that may support the contextual categorisation. For instance, it is possible that the creation of bonds to the discipline that I mention in connection with popularisation is connected to a higher degree of personalisation of the discipline. If the discipline is not just knowledge and facts, but actually represented by an (inviting) person, it may be easier to bond with the expertise and the discipline. In the same vein, as I have proposed the knowledge communication concept of citizen science as the category with the highest degree of participatory ambition, it could be interesting to investigate the textual differences between communicative instances from this category to see how the higher degree of ambition is visible in the texts, compared to texts from other categories. Hence, there is work to be done in order to put meat on the bones of the continuum of knowledge communication concepts proposed here.

References

Calsamiglia, H., & Van Dijk, T. A. (2004). Popularization discourse and knowledge about the genome. *Discourse & Society, 15*(4), 369–389. https://doi.org/10.1177/0957926504043705

Engberg, J. (2015). LSP studies as a quest for meso-level regularities. In G. Budin & V. Lušicky (Eds.), *Languages for special purposes in a multilingual, transcul-*

tural world, proceedings of the 19th European Symposium on Languages for Special Purposes, Keynote Addresses, 8–10 July 2013, Vienna, Austria (pp. 14–25). University of Vienna. https://lsp2013.univie.ac.at/proceedings-of-keynote-addresses/

Engberg, J. (2016). Conceptualising corporate criminal liability: Legal linguistics and the combination of descriptive lenses. In G. Tessuto, V. K. Bhatia, G. Garzone, R. Salvi, & C. Williams (Eds.), *Constructing legal discourses and social practices: Issues and perspectives* (pp. 28–56). Cambridge Scholars.

Engberg, J. (2020a). Institutional dissemination of legal knowledge—An instance of knowledge communication. In M. Gotti, S. Maci, & M. Sala (Eds.), *Scholarly pathways: Knowledge transfer and knowledge exchange in academia* (pp. 175–205). Peter Lang.

Engberg, J. (2020b). Multimodal institutional knowledge dissemination and popularization in an EU context—Explanatory ambition in focus. In G. Tessuto, V. K. Bhatia, R. Breeze, N. Brownlees, & M. Solly (Eds.), *The context and media of legal discourse* (pp. 50–76). Cambridge Scholars Publishing.

Engberg, J. (2022). LSP and transdiscursive knowledge communication. In E. Isaeva (Ed.), *Specialized knowledge mediation: Ontological & metaphorical modelling* (pp. 61–77). Springer International Publishing. https://doi.org/10.1007/978-3-030-95104-7_3

Engberg, J., Cacchiani, S., Luttermann, K., & Preite, C. (2018). Studying popularisation in legal communication: Introduction and overview. In J. Engberg, S. Cacchiani, K. Luttermann, & C. Preite (Eds.), *Popularization and knowledge mediation in the legal field* (pp. IX–XXV). LIT Verlag.

Engberg, J., & Heller, D. (2020). Rechtliches Wissen auf einer institutionellen Webseite. Eine Fallstudie zu sprach- und wissensorientierten Techniken der Popularisierung von Wissen im Netz. In K. Beckers & M. Wassermann (Eds.), *Wissenskommunikation im Web. Sprachwissenschaftliche Perspektiven und Analysen* (pp. 37–63). Peter Lang.

Engberg, J., Fage-Butler, A., & Kastberg, P. (Eds.). (2023). *Perspectives on knowledge communication: Concepts and settings*. Routledge.

Gotti, M. (2014). Reformulation and recontextualization in popularization discourse. *Ibérica, 27*, 15–34.

Habermas, J. (1981). *Theorie des Kommunikativen Handelns, Band 1*. Suhrkamp.

Jaki, S. (2018). Emotionalisierung in TV-Dokus zum Thema Jack the Ripper. Eine analyse mit DIMEAN. In L. Blasch, D. Pfurtscheller, & T. Schröder (Eds.), *Schneller, bunter, leichter. Kommunikationsstile im medialen Wandel* (pp. 213–231). iup.

Jaki, S. (2020). Zur Ausprägung neuer hybrider Dokuformate. Die Netflix-Serie 'Roman Empire' aus medienlinguistischer Perspektive. In M. Agnetta & M. Schleich (Eds.), *Folge um Folge. Multiple Perspektiven auf die Fernsehserie* (pp. 95–112). Universitätsverlag Hildesheim.

Kastberg, P. (2011). Knowledge asymmetries: Beyond to have and have not. *Fachsprache—International Journal of Specialized Communication, 34*(3–4), 137–151.

Kastberg, P. (2019). *Knowledge communication. Contours of a research agenda.* Frank & Timme. https://ebookcentral-proquest-com.ez.statsbiblioteket.dk:12048/lib/asb/detail.action?pq-origsite=primo&docID=5999996

Liebert, W.-A. (2002). *Wissenstransformationen: Handlungssemantische Analysen von Wissenschafts- und Vermittlungstexten.* de Gruyter.

Lorés, R. (2020). Science on the web: The exploration of European research websites of energy-related projects as digital genres for the promotion of values. *Discourse, Context & Media, 35*, 1–10. https://doi.org/10.1016/j.dcm.2020.100389

Luzón, M.-J., & Pérez-Llantada, C. (2019). Connecting traditional and new genres. Trends and emerging themes. In C. Pérez-Llantada & M.-J. Luzón (Eds.), *Science communication on the Internet. Old genres meet new genres* (pp. 1–18).

Porup Thomasen, U. (2015). *Exploring the communicative dimensions of knowledge-intensive innovation: An ethnographic insight into the innovation culture initiative of Novo Nordisk.* Department of Business Communication, Aarhus University. https://pure.au.dk/ws/files/86923107/Exploring_the_Communicative_Dimensions_of_Knowledge_Intensive_Innovation.pdf

Redder, A. (2002). Sprachliches Handeln in der Universität—das Einschätzen zum Beispiel. In A. Redder (Ed.), *"Effektiv studieren". Texte und Diskurse in der Universität* (pp. 5–28). Gilles & Francke.

Robinson, L. D., Cawthray, J. L., West, S. E., Bonn, A., & Ansine, J. (2018). Ten principles of citizen science. In S. Hecker, M. Haklay, A. Bowser, Z. Makuch, J. Vogel, & A. Bonn (Eds.), *Citizen science: Innovation in open science, society and policy* (pp. 27–40). UCL Press. https://doi.org/10.14324/111.9781787352339

8

Video Abstracts for Increasing Researcher Visibility

Olga Dontcheva-Navratilova

1 Introduction

The video abstract has emerged as a new genre of academic interaction over the last decade. It appeared in response to new opportunities offered by the internet multimodal digital environment to personalise research papers, share complex information visually and thus extend the reach of scientific research. Video abstracts can help researchers market themselves and their products in the highly competitive research marketplace (Cheek & Øby, 2019) by constructing the researchers' digital identities, adjusting the researchers' public image to audience expectations and persuading the audience to accept the image researchers claim for themselves. While video abstracts have not yet become a standard component of scholarly publications, journals in the hard sciences tend to see them as "a new initiative […] which will further increase the visibility of our authors and their work" (Reupert, 2017, p. 2). This seems to stem from the

O. Dontcheva-Navratilova (✉)
Masarykova univerzita, Brno, Czech Republic
e-mail: navratilova@ped.muni.cz

R. Plo-Alastrué, I. Corona (eds.), *Digital Scientific Communication*,
https://doi.org/10.1007/978-3-031-38207-9_8

potential of video abstracts to increase citation counts (Zong et al., 2019; Coccetta 2021, p. 305) and impact based on 'likes' on YouTube (Spicer, 2014, p. 4), thus promoting the researchers and their work in traditional scholarly and modern open-science contexts.

Defined as "a video presentation corresponding to a specific science research article, which typically communicates the background of a study, methods used, study results and potential implications through the use of images, audio, video clips, and text" (Spicer, 2014, p. 3), the video abstract is closely related to the research article. However, similarly to the printed abstract, it is a distinct genre with a different function, rhetorical structure and multimodal realisation (cf. Lorés, 2004). As a remediated genre (cf. Luzón & Pérez-Llantada, 2022), that is, a genre which has been refashioned and changed as a result of migration to the web (such as the graphic abstract or the video presentation), the video abstract recontextualises the printed abstract from a multimodal perspective via selection, arrangement, foregrounding and social repositioning (Bezemer & Kress, 2008). Thus, apart from its clearly promotional aspect, the video abstract represents a novel site of multimodal interaction between the researcher and the audience. Yet, while there is a large body of studies on printed abstracts (e.g. Bondi & Lorés-Sanz, 2014; Salager-Meyer, 1992; Santos, 1996), the genre of the video abstract is still under-researched. Recent studies have explored the rhetorical structure of video abstracts in medical journals (Plastina, 2017; Coccetta, 2021) and on the Cell Press platform (Liu, 2020), the use of metadiscourse in video abstracts in management journals (Cavalieri, 2020) and the potential impact of video abstracts hosted on YouTube and journal websites (Spicer, 2014). These studies have indicated that video abstracts tend to show signs of genre hybridisation (Bhatia, 2004) as a result of their migration to the web, that is, they mix features of the printed abstract and the conference presentation. The rhetorical structure and modes used in video abstracts vary across several criteria: the discipline, the platform they are hosted on and the manner of production (professional vs non-professional); this variation is further enhanced by a lack of clear author guidelines and divergence in instructions to authors (Liu, 2021). Thus, further research is needed to understand this hybrid genre and to estimate its potential to increase researcher visibility.

This study draws on the genre analysis (Swales, 1990) and multimodal analysis (Kress, 2010) perspectives to study video abstracts in the field of mathematics with the aim of identifying their rhetorical structure and exploring how the employed multimodal patterns of interaction between the researcher and the audience enhance the visibility of the researcher. The analysis proposes a typology of video abstracts according to the interplay between the audio and the video components involved, identifies the core, conventional and optional rhetorical moves of the genre and studies how metadiscourse markers conveying personality and direct address to the audience may impact the promotion of the author. In addition, the rhetorical structure and function of video abstracts will be compared to those of printed abstracts to highlight the similarities and differences between the two genres.

2 Data and Method

This study is based on a small specialised corpus comprising video and printed abstracts of 16 research articles in the field of mathematics published online in the *Journal of Number Theory* (*JNT*; Elsevier), which according to Spicer (2014) was the Elsevier journal with the highest rate of average views per video abstract (1519) in the period May 2008–May 2013. The choice of this journal was motivated by the lack of previous research on video abstracts in the field of mathematics as well as by the availability of the video abstracts on the journal's webpage (Elsevier). In addition, in contrast with the video abstracts analysed by Liu (2020, 2021), the *JNT* video abstracts are non-professional, that is, they are created by the scientists themselves, which means that decisions concerning the selection of rhetorical structure, modalities involved, language and visual means are made by the researchers and not modified by a media provider.

JNT publishes research in pure mathematics devoted primarily to the study of integers and integer-valued functions. As typical of the discourse of mathematics, the texts of *JNT* research articles are essentially multimodal as they comprise symbolic notations, equations and images (O'Halloran, 2015). Processing mathematical discourse is challenging not only because of the use of multimodal formulations, but also linguistically, as it expresses relational processes conveyed by dense noun phrases, conjunctions and implicit logical connections (Schleppegrell, 2007, p. 139). It is therefore only natural that mathematicians are prone to use different ways, including video abstracts, to make their research more accessible to a potentially larger audience.

The *JNT* abstract requirements state that authors should submit concise and factual printed abstracts (150–200 words) indicating the research purpose, principal results and major conclusions. Video abstracts are not required, but if submitted, they are made available on YouTube and the journal's web page. The corpus used in this study comprises all video abstracts published in *JNT* in 2016 and their corresponding printed abstracts. It includes ten multi-authored and six single-authored abstracts written by mathematicians from various linguacultural backgrounds (mostly American and Chinese, but also French, Argentinian, Brazilian and South African). The oral part of the video abstracts was transcribed to allow for a detailed textual analysis. The corpus size and composition are summarised in Table 8.1.

The interplay of modes in the discourse of mathematics (O'Halloran, 2015) and conference presentations (Morell, 2015) was taken into consideration. A taxonomy of modes employed in video abstracts emerged from the analysis of the *JNT* corpus. The following modes were discerned:

1) Verbal modes
   a) Spoken b) Written
2) Non-verbal modes
   a) Mathematical symbolism—mathematical notations and symbolic equations
   b) Mathematical images—diagrams and models

Table 8.1 Corpus size and composition

Abstracts	Publication	Number	Average length	Total size
Printed	2016	16	87.6 words (36–134)	1401 words
Video	2016	16	4.5 min (0.45–9.38)	72 min
Transcripts			654 words (103–1598)	10,464 words

c) Non-verbal visuals—setting, actions and processes
d) Author as protagonist—eye contact, facial expressions, gestures, body posture, clothes and grooming

It is assumed that the combination of modes may impact the genre structure as well as the interactive patterns in the video abstracts. Presentation patterns in video abstracts (broadly overlapping with Liu's (2021) move units, e.g. talking head, voice-over, press conference, mini-movie) serve as a basis for the typology of *JNT* video abstracts that this chapter undertakes to identify.

The analysis of the rhetorical structure of printed abstracts draws on the genre analysis framework (Swales, 1990); more specifically, it is based on Bhatia's (1994) four-move model structure comprising M1 *Introducing the purpose*, M2 *Describing the methodology*, M3 *Summarising the results* and M4 *Presenting the conclusions*. Since knowledge construction in pure mathematics is based on "substantiation of new results by proof and logical, mathematical reasoning" (Kuteeva & McGrath, 2015), which is reflected in the specific rhetorical structure of pure mathematics research articles (cf. Kuteeva & McGrath, 2015), this study suggests the following rhetorical structure for printed *JNT* abstracts: M1 *Introducing the problem*, M2 *Theory: definitions and calculations*, M3 *Providing proof* and M4 *Concluding the mathematical argument*. While clearly linked to the rhetorical structure of printed abstracts, video abstracts display greater move variability (Plastina, 2017; Cavalieri, 2020; Coccetta, 2021), as their multimodal character leads to genre hybridisation. Therefore, the analysis of the rhetorical structure of *JNT* video abstracts takes into consideration moves/steps typical of conference presentations, such as *Introducing the author*, *Introducing the title*, *Thanking the audience* and *Invitation to read the full paper* (cf. Rowley-Jolivet & Carter-Thomas, 2005; Cavalieri, 2020), which are categorised as socialising subphases by Coccetta (2021). The delimitation of obligatory, conventional and optional moves draws on Kanoksilapatham (2015), that is, a move is considered obligatory if found in 100 per cent, conventional if found in more than 60 per cent and optional if found in less than 60 per cent of the abstracts in the corpus.

Finally, the contribution of personality (the presence of the author and the projection of his/her authorial voice into the text) and audience engagement (the assumption of shared knowledge and direct appeal to the audience) to enhancement of researcher visibility is explored by a combination of analysis of the metadiscourse categories of self-mention and reader reference (Hyland, 2005) and multimodal discourse analysis, taking into consideration the affordances of the selected interaction pattern used in the video abstract (Luzón, 2019; Valeiras-Jurado, 2020).

3 Types of *JNT* Video Abstracts

A multimodal analysis has identified four types of presentation patterns reflecting the verbal and visual modes present in the *JNT* video abstracts, which represent a cline from a more visible to a less visible representation of the researchers in the video abstract.

P1: Talking head or body—The author (or one of the authors) of the paper is present on screen, viewed in close-up, speaking directly to the camera (audience). The posture, clothes and context in which the researcher is set indicate a specific interpersonal framework (e.g. formal and authoritative, casual and informal).

P2: Voice-over—The voice of the author, who is reading or commenting on a slideshow.

P3: Slideshow—A series of still images or pages of information (slides), which represents the abstract content.

P4: Captions and subtitles—Written texts included on screen for emphasising or explaining information, especially when it is of critical importance. In *JNT* video abstracts, these take either of the two forms: (i) subtitles highlighting the title of the paper, the name of the author(s) and their affiliation or (ii) written text on a white or blackboard presenting the main points/calculations or the whole text of the abstract. The latter type of captions is particularly useful for non-native speakers to increase intelligibility (Liu, 2020).

Based on the occurrence of these presentation patterns in the corpus, three basic types of *JNT* video abstracts were identified: conferential, lecturing and conversational video abstracts. As Table 8.2 shows, most of the abstracts in the corpus are of the conferential and lecturing type (each represented by seven occurrences), while only two represent the conversational type of video abstracts. Conferential video abstracts are the most extensive as their length is approximately 7 min (1193 words; i.e. approximately 10 times longer than a *JNT* printed abstract). Lecturing video abstracts tend to be the shortest, with an average duration of 1.25 min (191 words).

The conferential video abstract evokes a context of professional interaction between peers typical of conference presentations, as it comprises a slideshow with text including mathematical symbolism and the recorded voice of the researcher. As Fig. 8.1 shows, the initial slide indicates the title of the paper, the author(s) and their affiliation, and simultaneously

Table 8.2 Types of *JNT* video abstract in the corpus

Type	Presentation pattern		No
1a Conferential	Voice-over and slideshow	3	7
1b Personalised conferential	Talking head/body Voice-over and slideshow	4	
2 Lecturing	Talking head/body and caption		7
3 Conversational	Talking head/body		2

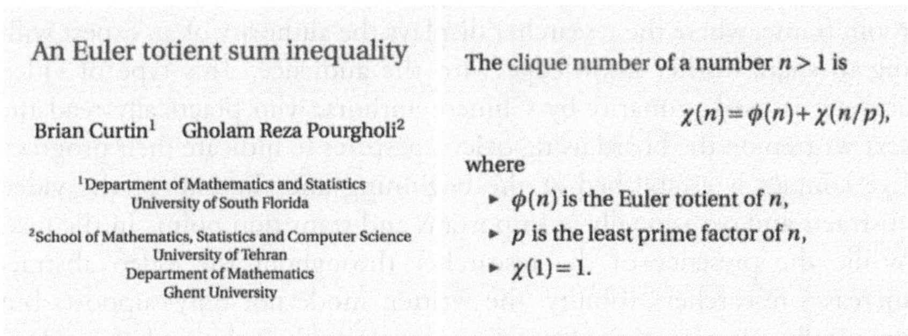

Fig. 8.1 The opening and content slides of a conferential video abstract (JNT_VA_02)

the recorded voice greets the audience and introduces the title and the authors, thus enhancing the visibility of the researcher. The sequence of slides that follows represents only the most important notions, symbolic equations and graphs discussed in the paper, while the voice of the researcher explains the details of the research results and calculations. Thus, the written mode in conferential video abstracts may be seen as supporting the spoken verbal mode (cf. Rowley-Jolivet, 2012) by reinforcing key words, utterances and mathematical symbols via the use of font sizes, colours, bulleting and other highlighting and outlining techniques, which can be considered written paralinguistic features (Morell, 2015, p. 141).

Personalised conferential abstracts open with a brief video (typically about 10 sec) showing the researcher in his/her office or in a non-formal context (e.g. sitting on a sofa, probably at home). The researcher maintains eye contact, uses interpersonal gestures and speaks directly to the camera to greet the audience and introduce the title and author(s) of the paper. This gives greater prominence to the researcher and sets the tone of the interaction on the scales of formality and social distance; for instance, video abstracts set in an informal setting are more likely to comprise personal narratives (e.g. personal involvement of the author with the topic, biographical details) and a higher rate of self-mention and reader reference devices (cf. Sect. 5).

The lecturing video abstract features the researcher in front of a black or white board displaying the text of the abstract including mathematical notations and equations. This setting activates the lecture-hall or classroom frame, where the researcher displays the authority of an expert willing to share his/her knowledge with the audience. This type of video abstract is used primarily by Chinese authors, who practically read the text written on the board using deictic gestures to indicate their progress. Eye contact is established at the beginning and the end of the video abstracts and occasionally at important and transition points in the text. While the presence of the researcher throughout the video abstract increases researcher visibility, the written mode not only supports but practically reiterates the information transmitted via the spoken mode.

Conversational video abstracts are clearly the most interactive type of video abstract, in which the content of the paper is transmitted exclusively via the spoken mode. This attributes a highly prominent role to the researcher, who imparts the message to the audience without the support of mathematical notations and equations. As Berkowitz (2013) notes, the option "to rethink one's research results in another format [...] is the immediate benefit of creating a video abstract". Moreover, conversational video abstracts tend to feature researchers in casual clothes and use informal settings, and to establish 'proximity' (Hyland 2010) with a wider audience (cf. Bell's 1984 audience design) by representing "not only themselves and their readers, but also their material, in ways which are most likely to meet their readers' expectations" (Hyland, 2010, p. 117). For instance, one of the conversational video abstracts in my corpus is set outdoors; it creates an informal friendly atmosphere by featuring the researcher sitting at a riverbank, smiling and waving his hand to greet the audience. The informal setting is complemented by the casual clothes of the researcher and a soundtrack with the noise of a river and the city. The researcher speaks to the camera while maintaining permanent eye contact and using interpersonal, cohesive and pragmatic gestures to engage with the audience and facilitate the processing of the argument chain. The only verbal written support for the viewer is a caption displaying the title of the paper, the name of the authors and their affiliation, which appear at the bottom of the video screen at the beginning and the end of the video abstract. This less technical more personal approach might explain why conversational abstracts seem to be more appealing to viewers, as they typically have a high view rate: 2177 and 533 for the two video abstracts in my corpus, as compared to the median of 395 views, ranging from 216 to 2177 (as of 30 May 2022).

The typology of *JNT* video abstracts indicates that researchers use various strategies to enhance their visibility and to address a wider audience. Apart from promoting the researcher by the presence of his/her image and voice, the different settings activating the conferential, lecturing and conversational frames help the researcher recontextualise the content of the paper by combining the affordances of the verbal and visual modes, thus involving the viewers in a familiar kind of interaction, establishing proximity and facilitating knowledge comprehension. All types of video abstracts extend the content of the respective printed abstract. The conferential video abstract extends the printed abstract significantly by including extensive argumentation and mathematical notations from the full article. In the case of the lecturing type, the extension is minimal (as suggested by their brevity) and commonly concerns the inclusion of mathematical notations (present in only ten of the printed abstracts) and elaboration on the research niche. The conversational video abstract represents a radical reformulation and refocusing of the paper by highlighting the motivation behind the research and the contribution to the field's knowledge.

4 Rhetorical Structure of *JNT* Video Abstracts

An analysis of the rhetorical structure of the *JNT* video abstracts shows that they include six moves; the first and the last move are interactive and comprise four and two steps, respectively, while the four central moves summarise the content of the paper.

The first move, M1 *Opening* (Example 1), is conventional, as it occurs in 14 *JNT* abstracts, that is, 87 per cent of all video abstracts (Table 8.3). It is realised by a voice-over in the case of conferential video abstracts and by a talking head/body in all other types of video abstracts and may include four steps:

Table 8.3 Move structure of *JNT* video abstracts in comparison with printed abstracts

JNT printed abstracts		*JNT* video abstracts		
		M1 Opening		14
		S1 Greeting the audience	*10*	
		S2 Introducing the author(s)	*11*	
		S3 Introducing the title	*14*	
		S4 Reference to JNT	*4*	
M1 Introducing the problem	16	M2 Introducing the problem		16
M2 Theory: calculations	8	M3 Theory: calculations		16
M3 Providing proof	16	M4 Providing proof		16
M4 Concluding the argument	6	M5 Concluding the argument		9
		M6 Closing		11
		S1 Thanking the audience	*10*	
		S2 Invitation to read the paper	*4*	

- S1 *Greeting the audience*
- S2 *Introducing the author(s) and their institutional affiliation*
- S3 *Introducing the title of the research paper*
- S4 *Reference to the journal in which the article is published*

Example (1), taken from a personalised conferential abstract, shows an instance of M1 comprising all four steps. The informal tone indicated by the greeting (the clothes and the home setting) is also visible in the transition to the slideshow, which is indicated verbally and visually by an explicit directive (*Let's go to the slides*) accompanied by a deictic gesture.

(1) Hello, my name is Joseph Vandehey. In this video I'll be giving an abstract for my upcoming paper 'New normality constructions for continued fractions', set to appear in the *Journal of Number Theory*. Let's go to the slides. (JNT_VA_05)

While S4 *Reference to the journal* is optional (as it occurs only four times in the corpus), the other three steps introducing the title, the author(s) and their affiliation are conventional (appearing 10, 11 and 14 times, respectively) and tend to be omitted only in personalised conferential abstracts, in which this information may be transmitted only via the written mode. As in conference presentations (cf. Carter-Thomas & Rowley-Jolivet, 2003), S1 *Greeting the audience* and S2 *Introducing the author(s)and their institutional affiliation* set the interpersonal framework by focusing on the researcher, appealing directly to the audience and indicating the contextual frame by visual cues, that is, slideshow, lecture hall, office, home or outdoor setting. Acknowledgement of co-authors, the researcher's affiliation and even S4 *Reference to the journal* may be interpreted as politeness strategies aimed at establishing in-group membership and claiming credibility and prestige as publication has been granted in a high-impact journal. The remaining step, S3, is more content oriented as it introduces the topic of the paper.

The second interpersonal move, M6 *Closing*, which indicates explicitly the end of the video, is realised by a talking head/body in lecturing and conversational video abstracts and by voice-over in both types of conferential video abstracts (Example 2). It is also conventional, as it appears in

68 per cent of the video abstracts (Table 8.3) and includes two steps, the first of which is conventional and the second optional:

- S1 *Thanking the audience*
- S2 *Inviting the audience to read the full paper*

(2) I invite you to read my article if you are interested in the details of the proof.
Thanks for watching. (JNT_VA_13)

Both steps appeal directly to the audience and may be seen as serving politeness strategies as they position the viewer as an interested colleague, that is, an in-group member, who shares the author's interest and expertise in the field.

The remaining four moves coincide with the moves of the printed abstract and summarise the content of the paper for the reader. As Table 8.3 shows, M2, M3 and M4, which present the basic stages of the development of a mathematical argument, are obligatory, as they occur in all *JNT* video abstracts. The first of these moves, M2 *Introducing the problem*, shows considerable variation as it may be an extensive move based on personal storytelling, a brief indication of research purpose by reference to previous research or citing of an existing problem, that is, retrieving a problem (Moghaddasi & Graves, 2017), as pure mathematics research prevailingly strives to solve existing statements or hypotheses which may have remained unresolved for decades or centuries (Example 3). However, in some cases, Move 2 may be embedded in M3 *Theory: definitions and calculations*, as definitions are often provided at the very beginning to clarify the argument and provide sufficient background to allow the audience to follow the proof, as in Example 4, where the problem is presented as a gap in previous knowledge. Move 3 typically comprises mathematical notations which provide the background to the study (cf. Kuteeva & McGrath, 2015).

(3) The main goal of this note is to show that this constant of the mixed (l_1, l_2)-Littlewood inequality is sharp. (JNT_VA_16)

(4) A sequence A is called d-complete if every sufficiently large integer is the sum of distinct terms taken from *A* such that no one divides the other. In 1996, Erdős and Lewin proved: the sequence {2α5β*p*γ} is *d*-complete for every prime *p* with $6 < p < 20$. They also commented that they cannot prove that the sequence {2α5β23γ} is *d*-complete. (JNT_VA_07)

M4, *Providing proof*, is a key move of the mathematics video abstract as it highlights the contribution of the paper. The proof may be presented in full by mathematical propositions, theorems or examples (typical of lecturing abstracts; Example 5), or it may simply be previewed in an effort to hook the audience and encourage them to read the whole paper (Examples 2 and 6).

(5) In fact, we proved the following main result: Let f be a multiplicative function and n_0 be a fixed integer with $1 \leq n_0 \leq 10^6$. If

(1) $f(p_0) \neq 0$ for at least one prime number p_0
(2) $f(p + q + n_0) = f(p) + f(q) + f(n_0)$ for all primes p, q
Then $f(n) = n$ for all integers n ≥1. (JNT_VA_08)

(6) In my paper, I show that the sequences that come from these non-linear recurrence sequences their Engel series also produced transcendental numbers. So, both comes by an application of Roth's theorem because it's possible to approximate them very well by rational numbers.

So, it's a reasonable guess that all of the Engel series of the kind I consider will produce transcendental numbers, but I don't know. (JNT_VA_11)

The only optional move is M5 *Concluding the mathematical argument*, which occurs in 56 per cent of the abstracts; this suggests that in pure mathematics providing the proof, that is, solving the problem, is often considered sufficient contribution to knowledge. Example 6 shows an interesting realisation of Move 5, in which the researcher hypothesises a possible implication of the study; however, the "I don't know" comment indicates that this is intended to indicate a direction for future research.

A comparison of the rhetorical structure of *JNT* video and printed abstracts (Table 8.3) indicates that only two of the obligatory moves in video abstracts have an obligatory status in printed abstracts, that is, M1 *Introducing the problem* and M3 *Providing proof*, while M2 *Theory: calculations* and M4 *Concluding the argument* are optional. This points to the high degree of condensation typical of the *JNT* printed abstracts, whose intended readership comprises experts in the field of number theory. It also highlights the more descriptive and explanatory character of video abstracts, which are intended for a potentially wider and more heterogeneous audience and therefore require a more careful presentation of the background to the study.

5 Personality and Audience Engagement in Printed and Video Abstracts

The occurrence of self-mention and reader reference in printed and video *JNT* abstracts is shown in Table 8.4. It is obvious that there is a major difference in the way printed and video abstracts use these interactional metadiscourse markers: the occurrence of self-mention is somewhat more frequent in printed abstracts (18.5 in printed abstracts vs 15.2 in video abstracts), while video abstracts are characterised by a very high rate of reader reference (16.8), a feature nearly absent in printed abstracts (0.7). The striking difference in the way video and printed abstracts employ reader reference is in consonance with Cavalieri's (2020) findings, who, following an exploration of management abstracts, reports a similar

Table 8.4 Comparison of self-mention and reader reference in printed and video *JNT* abstracts (per 1000 words)

Self-mentions and engagement markers	I/me/my	We/us/our exclusive	Total self-mention	We/us/our inclusive	You	One	Total engagement
Video abstracts	8.4	6.8	15.2	9.7	5.8	1.3	16.8
Printed abstracts	0	18.5	18.5	0	0	0.7	0.7

tendency. Audience involvement via the use of reader reference is generally typical of pure mathematics research articles, which, according to McGrath and Kuteeva (2012), show one of the highest rates of occurrence of this metadiscourse category (20.3 per 1000; see Hyland (2005) for a cross-disciplinary comparison). This may reflect efforts of mathematicians to position the reader/viewer as acquainted with the disciplinary conventions and apt to follow the argument chain that the researcher is willing to share with the interested audience.

An important difference is also present in the realisation of self-mention, which in printed abstracts is restricted to the exclusive *we*. In video abstracts, mathematicians also use *I* to enhance their involvement in the research process and cast a personal voice (a tendency also reported by Cavalieri (2020)). This increased employment of personality clearly aims to enhance the visibility of the researcher. There seems to be a functional specialisation in the use of plural and singular first-person pronouns expressing self-mention in video abstracts. Apart from the conventional use of *I/my* in M1 to introduce the researcher and the title of the paper (*I'm* + name, *my article/ paper* + title), the first-person singular pronoun is typically used for discourse organising purposes (*I'm going to start with, I want to tell you about, what I'm not going to do*), for defining or labelling concepts (*I'm going to call it, I'm going to denote*) and less frequently with research verbs to outline the premises or the proof (*I show, I prove, I consider, I do*) (Example 6). The exclusive *we*, on the other hand, is used only with research verbs (*we study, we show, we prove, we consider*) to comment on methods and to explain calculations. The possessive *our* indicates a strong association of the researchers with their finding, signalled by the collocations *our results, our formula* and *our theorem.*

Despite the presence of considerable idiosyncratic variation, a comparison of the use of self-mention and reader reference across the four types of video abstracts seems to indicate interesting tendencies (Table 8.5).

The personalised conferential abstract is clearly the video abstract type that makes most extensive use of self-mention (with *I* exceeding considerably *we* in number of occurrences, i.e. 4.5 vs 0.8) and reader reference (12.9). This seems to be related to the presence of the researcher at the beginning of the video abstract, which sets a more personalised

Table 8.5 Self-mention and reader reference across the four types of *JNT* video abstracts (per 1000 words)

Video abstract type	I/me/my	We/us/our exclusive	Total self-mention	We/us/our inclusive	You	One	Total reader reference
Conferential	0.9	3.4	4.3	0.4	1.4	0	1.8
Personalised conferential	4.5	0.8	5.3	8.9	2.8	1.2	12.9
Lecturing	0.8	2.2	3.0	0.3	0.8	0	1.1
Conversational	1.8	0.5	2.3	0	1.2	0.2	1.4

contextual frame, and to the considerable length of video abstracts using the audio mode in combination with the supporting visual mode. This allows researchers to get involved in storytelling, describe premises and proof in detail, share personal experience (Example 7) and engage with the audience by positioning them as in-group members when developing the argument chain (Example 8).

(7) So one thing that really drew me to the number theory as a graduate student was the interplay between arithmetic and analysis and I suppose one of the first places that I saw this interplay was in something like the class number formula for a billion number of fields. (JNT_VA_11)

(8) now as we look at the digits of π one could ask a very basic question do we see the number seven as often as we see the number four and do we see the string of digits 2 3 as often as we see the string of digits 0 8 if every string of digits is as common as every other string of digits of the same length then we say that π is normal in base 10. (JNT_VA_05)

The conferential video abstract does not include the visual presence of the researcher, which probably explains the lower rate of self-mentions and particularly of the singular first-person pronoun (0.9 for *I* vs 3.4 for *we*). Although the combination of the slideshow and the voice-over also provides sufficient space for explaining premises and proof, conferential video abstracts stay closer to the dominant stance of mathematics research articles, as they use mainly the exclusive *we* as a self-mention device

(Example 9). Reader reference is strikingly lower than in personalised conferential abstracts (1.8) and is confined primarily to the use of *you* for thanking the audience and inviting them to read the paper (Example 2).

(9) So actually we believe that any tower of GST type over a prime field has even genus but so far we have nothing to say about this question. (JNT_VA_02)

The low rate of self-mention and reader reference in lecturing abstracts seems to stem from their brevity. Although it features the researcher speaking to the camera in a lecture-hall setting, the lecturing video abstract is a kind of reading aloud of the printed abstract, which does not leave much space for establishing interpersonal relations with the audience; as a result, self-mention and reader reference occur primarily in the interactive opening and closing moves. The visibility of the researcher seems to be enhanced primarily by the visual rather than the verbal mode.

Rather surprisingly, the conversational abstract also employs a low rate of self-mention and reader reference markers (2.3 and 1.4, respectively). It seems that while establishing proximity via the visual mode (informal setting, clothes and soundtrack) and making extensive use of eye contact and pragmatic and interpersonal gestures, the verbal mode in conversational abstracts adopts a more distant stance. This may indicate an effort on the part of researchers to complement the more personal approach displayed by the visual mode with professional expertise cued by the verbal mode often associated with the use of nominalisation (Example 10).

(10) The motivation for this work was the realization that the positively shifted Mascheroni series are the dual sequence to Ramanujan sums of hyper harmonic numbers. (JNT_VA_09)

The variation in the conveyance of personality and audience engagement via the verbal mode across the four types of *JNT* video abstracts may be regarded as a cline reflecting the degree of visibility self-mentions and reader reference grant to the researcher, ranging from the personal conferential to the lecturing video abstract. However, meaning construction and the overall perception of researcher visibility in the genre of video abstracts seem to depend primarily on the interplay of various modes, and they cannot be confined to the affordance of a single mode.

6 Conclusion

This chapter has studied an innovative academic genre that offers new dissemination opportunities using the multimodal affordances of digital video presentation. The study has focused on an underexplored scientific discipline—pure mathematics, as represented in *JNT*—which tends to publish non-professional video abstracts. The results of the multimodal and genre analysis have shown that as a migrated genre, the video abstract recontextualises the printed abstract by refocusing and often extending its content by presenting the author(s), personal storytelling or commenting on mathematical equations to enhance the visibility of the researcher, establish proximity with the audience and facilitate knowledge comprehension.

The study has proposed a typology of video abstracts comprising three basic types according to the interplay between the oral and visual components involved: conferential (personalised and non-personalised), lecturing and conversational. The three types of video abstracts differ considerably in their length, modes used and the extent to which they recontextualise the printed abstract. While the lecturing abstract shows a low level of recontextualisation, the conferential abstract, especially the personalised sub-type, elaborates and refocuses the information provided in the printed abstract to assist the viewer in following the argument chain, and thus may reach a wider audience. Recontextualisation is particularly extensive in the conversational abstract, where the focus is shifted away from the presentation of proof to the motivation for the study and the contribution to knowledge.

The rhetorical structure of *JNT* video abstracts has been found to include six moves. The opening and the closing move, which comprise several steps and are similar to those in conference presentations (cf. Rowley-Jolivet & Carter-Thomas, 2005), are conventional; they are highly interactive and may be regarded as typical of video abstracts across scientific disciplines. The central four moves summarise the content of the paper and often show disciplinary variation. In pure mathematics, three of these moves are obligatory. The focus is on introducing the problem and presenting calculations and proof, while the 'Concluding the argument' move is optional, as in pure mathematics the proof is regarded as the main contribution to knowledge.

The potential of video abstracts to enhance researcher visibility and reach a wider audience is also indicated by more extensive use of metadiscourse to convey personality and audience engagement than is the case with printed abstracts. While in printed abstracts self-mention is confined to the exclusive *we* and there is a lack of reader pronouns, in video abstracts researchers tend to use frequently the whole spectrum of self-mention and reader reference devices (cf. Cavalieri, 2020). Self-mention structures are typically employed to organise the discourse, label concepts and show the agentive role of the researcher in the research process. Reader reference aims at creating communality with the audience by positioning the viewer as an interested peer and in-group member who follows the argument chain.

To conclude, while the video abstract has not yet become a firmly established genre of academic interaction, it clearly shows strong potential to enhance researcher visibility and disseminate scientific knowledge to wider audiences thanks to the affordances of multimodal digital technologies. However, since video abstracts show considerable variation in modes, presentation patterns, layout and duration due to a lack of clear guidelines, disciplinary differences and various modes of production, more research is needed to map the distinctive features of the genre across different disciplines and cultural contexts.

References

Bell, A. (1984). Language style as audience design. *Language in Society, 13*(2), 145–204.

Berkowitz, J. (2013, February 6). Video abstracts, the latest trend in scientific publishing: Will "publish or perish" soon include "video or vanish"? University Affairs. https://www.universityaffairs.ca/features/feature-article/video-abstracts-the-latest-trend-in-scientific-publishing/

Bezemer, J., & Kress, G. (2008). Writing in multimodal texts: A social semiotic account of designs for learning. *Written Communication, 25*, 166–195. https://doi.org/10.1177/0741088307313177

Bhatia, V. J. K. (1994). *Analysing genre: Language use in professional settings*. Routledge.

Bhatia, V. J. K. (2004). *Worlds of written discourse. A genre-based view*. Continuum.

Bondi, M., & Lorés-Sanz, R. (Eds.). (2014). *Abstracts in academic writing*. Peter Lang.

Carter-Thomas, S., & Rowley-Jolivet, E. (2003). Analysing the scientific conference presentation (CP). A methodological overview of a multimodal genre. *ASp, La Revue du GERAS, 39–40*, 59–72. https://doi.org/10.4000/asp.1295

Cavalieri, S. (2020). From abstracts to video-abstracts: Academic argumentation in genre hybridization. *Token: A Journal of English Linguistics, 11*, 41–58. https://doi.org/10.25951/4306

Cheek, J., & Øby, E. (2019). "Getting attention" creating and presenting the visible, online, researcher self. *Qualitative Inquiry, 25*(6), 571–582. https://doi.org/10.1177/1077800418806593

Coccetta, F. (2021). Medical video abstracts and their subgenres: A phase-based approach to the detection of generic structure patterns. *European Journal of English Studies, 25*(3), 316–333. https://doi.org/10.1080/13825577.2021.1988264

Hyland, K. (2005). *Metadiscourse: Exploring interaction in writing*. Continuum. https://doi.org/10.5040/9781350063617

Hyland, K. (2010). Constructing proximity: Relating to readers in popular and professional science. *Journal of English for Academic Purposes, 9*, 116–127. https://doi.org/10.1016/j.jeap.2010.02.003

Kanoksilapatham, B. (2015). Distinguishing textual features characterizing structural variation in research articles across three engineering sub-discipline corpora. *English for Specific Purposes, 37*, 74–86. https://doi.org/10.1016/j.esp.2014.06.008

Kress, G. (2010). *Multimodality: A social semiotic approach to contemporary communication*. Routledge.

Kuteeva, M., & McGrath, L. (2015). The theoretical research articles as a reflection of a disciplinary practice: The case of pure mathematics. *Applied Linguistics, 36*(2), 215–235. https://doi.org/10.1093/applin/amt042

Liu, J. (2020). Research video abstracts in the making: A revised move analysis. *Journal of Technical Writing and Communication 2020, 50*(4), 423–446. https://doi.org/10.1177/0047281619894981

Liu, J. (2021). Video or perish? An analysis of video abstract author guidelines. *Journal of Librarianship and Information Science*, 1–9. https://doi.org/10.1177/09610006211006774

Lorés, R. (2004). On RA abstracts: From rhetorical structure to thematic organisation. *English for Specific Purposes, 23*, 280–302. https://doi.org/10.1016/j.esp.2003.06.001

Luzón, M. J. (2019). Bridging the gap between experts and publics: The role of multimodality in disseminating research in online videos. *Iberica, 37*, 167–192.

Luzón, M.-J. & Pérez-Llantada, C. (2022). *Digital genres in academic knowledge production and communication: Perspectives and practices.* Multilingual Matters. https://doi.org/10.21832/9781788924726

McGrath, L., & Kuteeva, M. (2012). Stance and engagement in pure mathematics research articles: Linking discourse features to disciplinary practices. *English for Specific Purposes, 31*(3), 161–173. https://doi.org/10.1016/j.esp.2011.11.002

Moghaddasi, S., & Graves, H. (2017). "Since Hadwiger's conjecture is still open": Establishing a niche for research in discrete mathematics research article introductions. *English for Specific Purposes, 45*, 69–85. https://doi.org/10.1016/j.esp.2016.09.003

Morell, T. (2015). International conference paper presentations: A multimodal analysis to determine effectiveness. *English for Specific Purposes, 37*, 137–150. https://doi.org/10.1016/j.esp.2014.10.002

O'Halloran, K. (2015). The language of learning mathematics: A multimodal perspective. *Journal of Mathematical Behavior, 40*, 63–74. https://doi.org/10.1016/j.jmathb.2014.09.002

Plastina, A. F. (2017). Professional discourse in video abstracts: Re-articulating the meaning of written research article abstracts. In G. Garzone, P. Catenaccio, K. Grego, & R. Doerr (Eds.), *Specialised and professional discourse across media and genres* (pp. 57–74). Ledizioni.

Reupert, A. (2017). Promoting research through video abstracts (Editorial). *Advances in Mental Health, 15*(1), 1–3. https://doi.org/10.1080/18387357.2017.1312761

Rowley-Jolivet, E. (2012). Oralising text slides in scientific conference presentations: A multimodal corpus analysis. In A. Boulton, S. Carter-Thomas, &

E. Rowley-Jolivet (Eds.), *Corpus-informed research and learning in ESP: Issues and applications* (pp. 137–166). https://doi.org/10.1075/scl.52.06row

Rowley-Jolivet, E., & Carter-Thomas, S. (2005). The rhetoric of conference presentation introductions: Context, argument and interaction. *International Journal of Applied Linguistics, 15*(1), 45–70. https://doi.org/10.1111/j.1473-4192.2005.00080.x

Salager-Meyer, F. (1992). A text-type and move analysis study of verb tense and modality distribution in medical English abstracts. *English for Specific Purposes, 11*, 93–113. https://doi.org/10.1016/S0889-4906(05)80002-X

Santos, M. (1996). The text organization of research papers abstracts in applied linguistics. *Text, 16*, 481–499. https://doi.org/10.1515/text.1.1996.16.4.481

Schleppegrell, M. J. (2007). The linguistic challenges of mathematics teaching and learning: A research review. *Reading and Writing Quarterly, 23*(2), 139–159. https://doi.org/10.1080/10573560601158461

Spicer, S. (2014). Exploring video abstracts in science journals: An overview and case study. *Journal of Librarianship and Scholarly Communication, 2*(2), eP1110.

Swales, J. (1990). *Genre analysis. English in academic and research settings*. Cambridge University Press.

Valeiras-Jurado, J. (2020). Genre-specific persuasion in oral presentations: Adaptation to the audience through multimodal persuasive strategies. *International Journal of Applied Linguistics, 30*, 293–312. https://doi.org/10.1111/ijal.12284

Zong, Q., Xie, Y., Tuo, R., Huang, J., & Yang, Y. (2019). The impact of video abstract on citation counts: Evidence from a retrospective cohort study of New Journal of Physics. *Scientometrics, 119*(3), 1715–1727. https://doi.org/10.1007/s11192-019-03108-w

9

Reconceptualisation of Genre(s) in Scholarly and Scientific Digital Practices: A Look at Multimodal Online Genres for the Dissemination of Science

Noelia Ruiz-Madrid and Julia Valeiras-Jurado

1 Introduction

The ways in which we communicate science have experienced a rapid evolution in the past decades. As Scotto di Carlo (2014, 2015) points out, the emergence of new communication channels has brought science closer to the public without an academic institution acting as an intermediary (see Engberg in this volume, for a deeper discussion). Among these new communication channels, the internet enjoys a privileged position. Previous studies by Luzón and Pérez-Llantada (2019), Pérez-Llantada (2021), Rowley-Jolivet and Carter-Thomas (2019, 2020) or Carter-Thomas and Rowley-Jolivet (2020) claim that it has even changed the rules of scientific communication, maximising the possibilities to reach a broader audience, blurring the boundaries between the scientific

---

N. Ruiz-Madrid (✉) • J. Valeiras-Jurado
Universitat Jaume I, Castelló, Spain
e-mail: madrid@uji.es; valeiras@uji.es

R. Plo-Alastrué, I. Corona (eds.), *Digital Scientific Communication*,
https://doi.org/10.1007/978-3-031-38207-9_9

community and the general public, and forcing a process of adaptation to this new and diversified audience. Digital genres, thus, seem to be the natural means for open science communication, since they afford transparency, accessibility, sharedness and collaboratively developed knowledge in science communication, which, according to Vicente-Saéz and Martínez-Sanz (2018), are the true features of open science. Open science and digital genres seem to go hand in hand to the point that the Organisation for Economic Co-operation and Development (OECD, 2015, p. 7) defines open science as making "the primary outputs of publicly funded research results—publications and the research data—publicly accessible in digital format with no or minimal restriction," and as put forward by Luzón and Pérez-Llantada (2022, p. 31) "although knowledge communication is not always open, the internet makes it easier to share ideas, data and results, thus enabling open science."

In this chapter, we want to focus our attention on one internet-based genre that has gained increasing importance as a tool to communicate science to a broader audience: popular science online videos. These are short, professionally produced videos of approximately 10 minutes, in which a varied and validated scientific content is disseminated to the general public. As pointed out by Valeiras-Jurado and Bernad-Mechó (2022), they are a relatively recent phenomenon and they have not yet been defined as a genre, but it can be safely claimed that they share characteristics of what some scholars have called scifotainment or edutainment genres (Pérez-Llantada, 2021; Sancho Guinda, 2019; Zhang & O'Halloran, 2013).

A salient characteristic of these videos is their multimodal nature, which has been pointed out in previous research. Boy et al. (2020) describe them as

> well-organized multimodal arrangements consisting of a variety of visual and verbal modes like stills, moving images, text, spoken language, sounds, animations, graphics, etc. which is a much more complex system of communication than text only. (p. 4)

These authors have classified scientific online videos according to their use of semiotic modes. They distinguish between:

1. *Presentation films*, in which the speaker talks directly to the camera and answers some scientific questions. In addition to spoken language, they use other modes such as text over visuals, background images, animations, gestures and facial expressions.
2. *Expert films*, in which an expert discusses a topical field of research. These videos contain more animations, insertion of user comments, addresses to users and so on.
3. *Animation films* show artificial moving images to illustrate a scientific issue.
4. *Narrative explanatory films* are based on a general question that is answered through arguments that follow logical reasoning. They often contain elements from the previous types, combining narrative and informative elements in an entertaining story. They use mainly moving image material.

According to Valeiras-Jurado and Bernad-Mechó (2022), something that distinguishes these videos is that they combine modes that are orchestrated by presenters as they speak to the camera, which can be classified as embodied modes, and other modes that are added to the ensemble during montage, which can be called filmic modes. This is a possibility afforded by the online media and turns meaning-making in these videos into a very complex and rich process. It also brings to the fore the importance of co-authorship in this genre: in popular science online videos, multimodal ensembles are collaboratively orchestrated by the presenter and the person in charge of montage (Xia & Hafner, 2021). This peculiarity brings with it the extra challenge of ensuring the internal coherence of the ensemble, by making sure that all modes contribute seamlessly towards the same communicative aim (Valeiras-Jurado, 2019; Valeiras-Jurado & Bernad-Mechó, 2022).

The process of meaning-making in popular science online videos is further challenged by the need to account for the knowledge asymmetry between presenters and audience, that is, presenters have considerably more knowledge than the audience about the topic they are discussing (Kastberg, 2011). Therefore, scientific content needs to be adapted by means of a process of recontextualisation (Luzón & Pérez-Llantada, 2019). According to Bezemer and Kress (2008), recontextualisation involves four rhetorical principles:

(a) *Selection* of the content that is relevant for the new audience
(b) *Arrangement* of the content in the way that best fits the audience and the communicative purpose
(c) *Foregrounding* of the elements that are particularly significant
(d) *Social reposition* that is redefining the relationship between the speaker/writer/presenter and the audience

Previous studies have investigated the way these principles are put into practice, with special focus on academic and scientific discourse. Different taxonomies of recontextualisation strategies have been proposed. Luzón (2019), for example, identifies four groups of strategies depending on their function: (a) strategies to build credibility, (b) strategies to build persuasive arguments, (c) strategies to tailor information to the assumed knowledge of the audience and (d) strategies to engage the audience. She also approaches these recontextualisation strategies from a multimodal perspective, acknowledging the role of static and moving image, among others.

Carter-Thomas and Rowley-Jolivet (2020) have investigated two of the groups proposed by Luzón (2019) in Three-Minute Theses presentations: strategies to tailor information and engagement strategies. Among the first they include selecting content, definitions, explanations in layman's language, paraphrases, analogies, examples from everyday life and scenarios (i.e., the audience is asked to imagine a particular situation). Among the second, they report the use of catchy titles, visual impact (e.g., through striking images or the presenter's presence on stage), various personalisation devices (e.g., gestures, smiles, personal pronouns), questions, humour and "street cred" (i.e., a common framework based on shared cultural values rather than scientific know-how).

Valeiras-Jurado et al. (2018) suggest a taxonomy of persuasive strategies found in a corpus of conference presentations and research dissemination talks. They identify strategies such as anticipation and control of responses (i.e., predicting potential responses from the audience and prompting desirable ones), attention getting (i.e., raising and maintaining the interest of the audience), emphasis (i.e., highlighting parts of the message to make them more salient), evaluation (i.e., assessing something and implicitly inviting the listener to accept this assessment), processing aids (facilitating understanding of the message), projection of the context of interaction (i.e., presenting information as new vs. given; or agreed upon vs open to discussion) and rapport (i.e., building a relationship of sympathy and mutual understanding with the audience).

To bring together the two pillars of our theoretical framework, that is, recontextualisation and multimodality, Bezemer and Kress define recontextualisation from a multimodal perspective as follows:

> moving meaning material from one context with its social organization of participants and its modal ensembles to another, with its different social organization and modal ensembles. Meaning material always has a semiotic realization, so recontextualization involves the re-presentation of the meaning materials in a manner apt for the new context in the light of the available modal resources. (2008, p. 184)

In the case of popular science videos that concern our study, recontextualisation implies moving scientific content to a novel online context in which new combinations of semiotic modes (i.e., multimodal ensembles) become available in order to meet the audience's needs and expectations, which are no longer the same as those from experts.

In this chapter it is our intention to unveil multimodal recontextualisation practices in popular science videos. In particular, we pose the following research questions:

- RQ1: What modes and modal configurations can be identified in popular science videos?
- RQ2: What recontextualisation strategies are used in these videos?
- RQ3: How are these strategies realised multimodally?

2 Corpus and Method

Multimodal ensembles refer to the orchestration of different modes to produce a specific meaning that is inferred from the interrelation among the specific modes involved. The communicator orchestrates multimodal ensembles, where each mode has a function (Kress, 2010) and "each mode is partial in relation to the whole meaning" (Jewitt & Kress, 2003, p. 3). In order to explore the nature of the multimodal ensembles employed by speakers to recontextualise information in the online science dissemination videos, we follow a multimodal analysis approach. A more detailed account of this methodology can be found in Sect. 2.2.

2.1 The Corpus

Four science dissemination online videos available on YouTube were selected for the present study: (a) *Why Megalodon (Definitely) Went Extinct*, (b) *Why Are We the Only Humans Left?*, (c) *Why the Muon g-2 Results Are So Exciting!* and (d) *What Is the Magic Russian Diamond?* Videos (a) and (b) belong to the anthropology field and videos (c) and (d) deal with physics. All of them were hosted in the PBS multi-platform media (https://www.pbs.org/) at the time the analysis was carried out. PBS is a private, non-profit corporation, which defines itself as "a trusted window to the world" and "America's largest classroom" that offers programming for a wide range of ages, interests and genres about science, history, culture, great literature and public affairs. The contents broadcast are aimed at a non-expert audience who wants to be updated or teachers who may use the contents for pedagogical purposes. The first video, *Why Megalodon (Definitely) Went Extinct*, has a duration of 11 min 12 sec and it is the seventh episode of season 2 of the channel *Eons*. This channel offers episodes on the history of life on Earth from the Archean Eon through the Mesozoic Era, that is, from the so-called Age of Dinosaurs to the end of the most recent Ice Age. This particular episode is narrated by actress and science communicator Kallie Moore and describes the evolution of the ancient species, particularly the Megalodon and its eventual extinction. The narrator discusses the fossil evidence and the effect the extinction of the Megalodon had on the ocean habitat and the current species. The second video, *Why Are We the Only Humans Left?* is 7 min 31 sec and it is the 30th episode of season 4 of the channel *Be Smart* presented by Joe Hanson, a science writer, biologist and YouTube educator.

This particular episode investigates human ancestry and explains why *Homo sapiens* are the only survivors on the human evolutionary tree and offers a new view of the Neanderthals. The third video, *Why the Muon g-2 Results Are So Exciting*, is the longest of the videos selected with a length of 12 min 35 sec. It is the 13th episode of season 7 of the channel *Space Time*, which deals with space and astrophysics. This channel is presented by Mathew O'Dowd, an astrophysicist who studies black hole physics and received his PhD from Columbia University. The chapter we analyse explains the results obtained by one of the most promising experiments in the world of physics: Muon G-2. Finally, the fourth video, *What Is the Magic Russian Diamond?* has a duration of 8 min 27 sec and it is episode 10 of season 3 of the channel called *Physics girl*. This channel is presented by science communicator Dianna Cowern. It explores the field of physical sciences and presents experiments, demonstrations and new discoveries. The episode analysed in the present paper describes the properties and applications of one of the most unusual diamonds in the world because of its properties and composition, that is, the Russian diamond.

The aim of the four videos is the dissemination of scientific content to a non-expert audience and mainly for pedagogical purposes. Therefore, these videos may resort to different semiotic resources and rhetorical strategies to recontextualise the scientific content and eventually achieve their communicative purpose.

The total size of the corpus is 39 m 45 s, which does not allow for quantitatively based generalisations, but it is valid for qualitative analyses considering the duration of the videos analysed (i.e., no longer than 13 min). It is also in line with similar previous multimodal studies (Luzón, 2019; Valeiras-Jurado & Ruiz-Madrid, 2019; Ruiz-Madrid, 2021; Valeiras-Jurado & Bernad-Mechó, 2022), which cannot afford to use larger corpora due to their minute level of data analysis and the lack of automating tools. Table 9.1 provides an overview of the corpus.

As Table 9.1 shows, the shortest video was 7 m 31 s and the longest one was 12 m 35 s. The corpus was varied in terms of disciplines: two videos from anthropology and two from physics. All of them were professionally edited and employed a wide variety of semiotic and filmic modes (Valeiras-Jurado & Bernad-Mechó, 2022).

Table 9.1 Description of the corpus

Video	Topic	Discipline	Speaker	Duration
Why Megalodon (Definitely) Went Extinct	Megalodon/ Sharks	Anthropology	Male	11 min 12 sec
Why Are We the Only Humans Left?	Prehistoric human beings	Anthropology	Female	7 min 31 sec
Why the Muon g-2 Results Are So Exciting!	Particle Physics	Physics	Male	12 min 35 sec
What Is the Magic Russian Diamond?	Composition of diamonds	Physics	Female	8 min 27 sec

2.2 Method and Instruments

In this research, a multimodal discourse analysis (MDA) approach (Luzón, 2019; Ruiz-Madrid, 2021; Valeiras-Jurado & Bernad-Mechó, 2022) is adopted to analyse and gain further understanding on how recontextualisation is multimodally conveyed in online science videos. From this perspective, we focus on how the different semiotic modes are employed in these videos and more precisely in the multimodal ensembles resulting when recontextualising the information. As noted by Paltridge (2012), MDA results in too much data for a feasible analysis, and therefore, it is necessary to select specific modes to look into. Accordingly, we made a first selection of modes informed by previous studies (Querol-Julián, 2011; Valeiras-Jurado & Ruiz-Madrid, 2019), which can be summarised as follows:

(a) Words (written and oral discourse), mainly those used as recontextualisation strategies (Rowley-Jolivet & Carter-Thomas, 2019; Luzón, 2019; Valeiras-Jurado et al., 2018)
(b) Four types of kinesics: gestures, head movement, facial expression (Kendon, 2004) and gaze and eye contact (Argyle et al., 1981) as well as the way they interact with speech (Ladewig, 2013; Müller, 2013)
(c) Image, music, sound and visual effects (Valeiras-Jurado & Bernad-Mechó, 2022)

This list was completed with more modes suggested by the corpus-driven analysis such as graphs and equations, clothing or artefacts, adopting an inductive approach to video data (Goldman et al., 2007). The final list of modes that were employed for the multimodal analysis is the following: spoken words, gestures, head movements, face expression, eye contact, image, written words, graphs and equations, video, music, sound effects, visual effects, clothing and artefacts and physical objects.

Concerning the recontextualisation strategies, we again used a corpus-driven approach that was informed by previous studies (Rowley-Jolivet & Carter-Thomas, 2019; Luzón, 2019; Valeiras-Jurado et al., 2018). Accordingly, nineteen strategies suggested in these previous studies were considered and classified into three different groups according to their rhetorical aim, namely, (a) strategies to tailor information, (b) strategies to engage the audience and (c) strategies to build credibility (see Sect. 1 for a full discussion). A detailed description of the recontextualisation strategies found in the present study can be found in Sect. 3.

A computer-aided multimodal analysis of the four videos was conducted with the software MMAV (Multimodal Analysis Video) (O'Halloran et al., 2015). This multimodal annotation tool makes it possible to transcribe and annotate audio and video files and to organise transcriptions and annotations in different layers. For the multimodal analysis the layer structure shown in Fig. 9.1 was used:

As shown in Fig. 9.1, the software is organised into three components: a set of media files, a set of categorical descriptions (i.e., systems) used in the annotation and a set of annotation units (with time-stamped and spatial coordinates). The software provides access to plain text, images, sound and videos, which cover to a major extent the ways multimodal phenomena can be digitally recorded. We imported the media file and designed a set of annotation systems with the list of modes mentioned above. As the corpus-driven analysis of the different videos was carried out, this set of annotation systems was modified accordingly. The MMAV offers a "state machine" tool that affords the visualisation of the annotation results as well as their quantification and the relationship among the created systems in a visual way, as shown in Fig. 9.2.

The videos were individually visualised and annotated by both researchers in order to ensure validity. Most disagreements concerned the

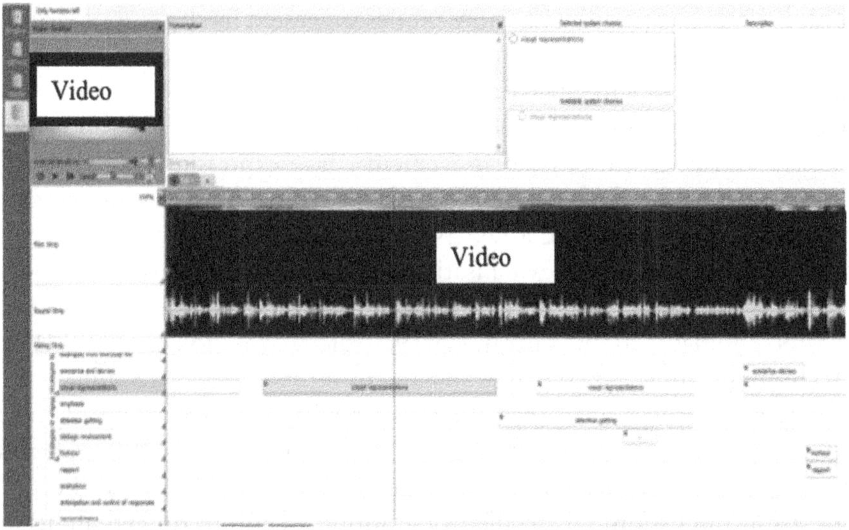

Fig. 9.1 Screenshot of the analysis of *Why Are We the Only Humans Left?* in MMAV

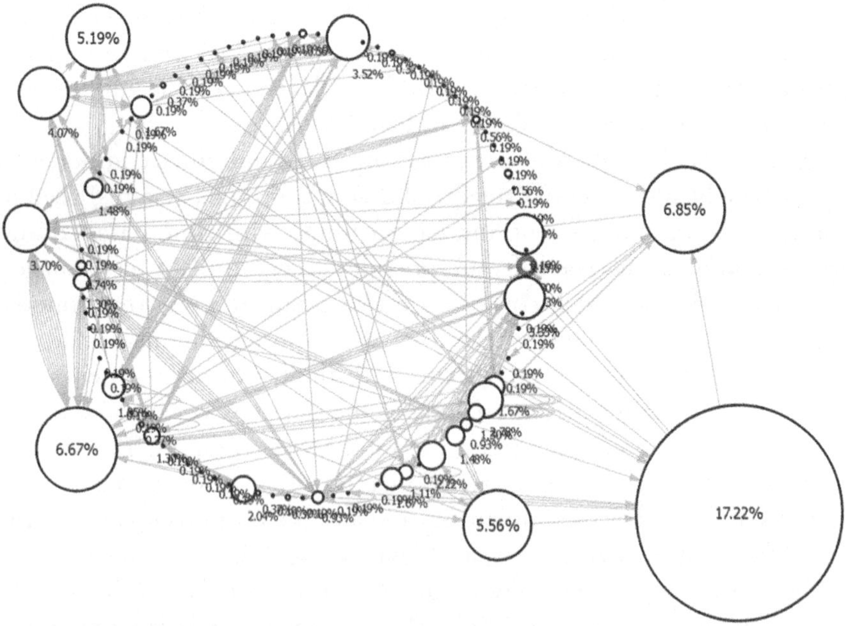

Fig. 9.2 Screenshot of the visualisation of results in *Why Are We the Only Humans Left?* provided by the state machine tool

annotation of the type of strategies. Consensus was reached by taking into account the communicative aim of the whole multimodal ensemble. In addition, it was acknowledged that modes can frequently have several functions at the same time and that the boundaries are never clear-cut.

3 Results and Discussion

In this section, we present and discuss which multimodal ensembles are employed to recontextualise information in online videos for science dissemination.

Regarding RQ1, that is, the modes and modal configurations that can be identified in popular science videos, we adopted an inductive approach to video data (Goldman et al., 2007) to identify the semiotic modes that contribute to conveying meaning, as explained in Sect. 2. The identification was made first individually, and the few inconsistencies between researchers discussed. The agreed list of modes, which was used as the basis for our coding library, is shown in Table 9.2.

Table 9.2 Modes used in the coding library

Mode	Instances from the corpus
Spoken words	That is, what the presenter says
Written words	That is, text that appears on screen
Gestures	That is, movements of the hand and arms
Head movements	For example, tilt, shake, nod
Face expression	For example, frown, smile
Eye contact	For example, gazing at the camera or somewhere else
Graphs and equations	For example, a chart, Einstein's equation $E = mc^2$
Image	Including real (e.g., photographs of people, animals and objects) and cartoons or depictions
Video	Including real and animated
Music	It is frequently present at the beginning and end of the videos, and is also hearable quieter, as background, during the video
Sound effects	For example, an alarm, a typing sound
Visual effects	For example, movements in images and in the position of the presenter
Clothing	The clothes worn by the presenter
Artefacts	Physical objects manipulated in the videos

The use of *video* as one of the modes requires some further explanation, since the communicative events we are analysing are also videos in themselves. We consider *video* as a mode (or rather a collection of modes including other modes like image, sound and music) when it is pre-recorded and embedded in the episode (e.g., an interview with a scientist in *What Is the Magic Russian Diamond?*). It is a combination of semiotic resources that is added to the episode during montage, and, unlike gestures or face expression, not controlled by the presenter in the video. Together with music, sound effects and visual effects, they are part of what Valeiras-Jurado and Bernad-Mechó (2022) call filmic modes. Likewise, the last two modes in Table 9.2 (i.e., *clothing* and *artefacts*) may not be so straightforward as a way of conveying meaning. We consider that clothes become a mode when they are worn intentionally. This is the case, for example, of a T-shirt shown in *Why the Muon g-2 Results Are So Exciting!* displaying the names and logos of PBS studios and the name of the channel, or a lab coat used in *What Is the Magic Russian Diamond?* to engage in scientist lecture mode, as Fig. 9.3 shows.

Artefacts are physical objects that are being manipulated. They also become a mode, especially when they are used to visually illustrate what

Fig. 9.3 Examples of clothing conveying different strategies in *Why the Muon g-2 Results Are So Exciting!* and *What Is the Magic Russian Diamond?*

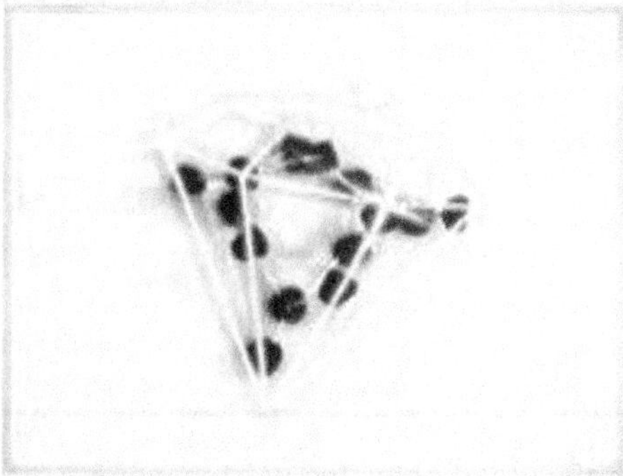

Fig. 9.4 Example of the use of artefacts for visualisation purposes in *What Is the Magic Russian Diamond?*

the presenter is explaining. Figure 9.4 shows the use of gummies to visually represent the structure of a diamond in *What Is the Magic Russian Diamond?*

The multimodal annotation revealed that modes are used with great flexibility and creativity, and as a result, it was not possible to identify clearly recurrent multimodal ensembles. However, we were able to identify modes that tended to appear together. For example, *spoken words* almost always tend to co-occur with *gestures* whenever the presenter is visualised on screen. Another frequent combination is *image* and *visual effects*. In fact, we noticed that these videos avoid static images and prefer images that move or change in some way. Also text, which is used less frequently than images, tends to be animated in some way, for example, with a pop-up effect, disappearing from screen and a special typography. These semiotic resources, which for the purposes of this study we have gathered under the filmic mode *visual effects* (rather than as part of *written words*, since they also affect images), have been shown to contribute to the multimodal ensemble in previous studies (Stöckl, 2014; van Leeuwen, 2006). We believe they provide the videos with greater dynamism and help retain viewers' attention. Figure 9.5 illustrates this trend.

We also noticed a certain regularity in the sequencing of modes. We found two recurrent sequences in our corpus:

Fig. 9.5 Example of the use of animated written words

Fig. 9.6 Sequence in *Why the Muon g-2 Results Are So Exciting!*

(1) A visualisation of the presenter speaking leads to an *image*, *text* or *graph* appearing in the background. This image stays static very briefly, and then a *visual effect* is applied, which makes the *image* move, change size, disappear and so on.

Figure 9.6 shows this sequence in the *Muon* video. The ensemble begins with the *speaker* talking to the camera and using *gestures* as he speaks, and then an *image* is shown in the background. This image stays static very briefly and then movement is introduced through *visual effects* (in this case the balance tips).

Another example from the *Megalodon* video is provided in Fig. 9.7, including the annotations of the modes. As in the previous case, we see

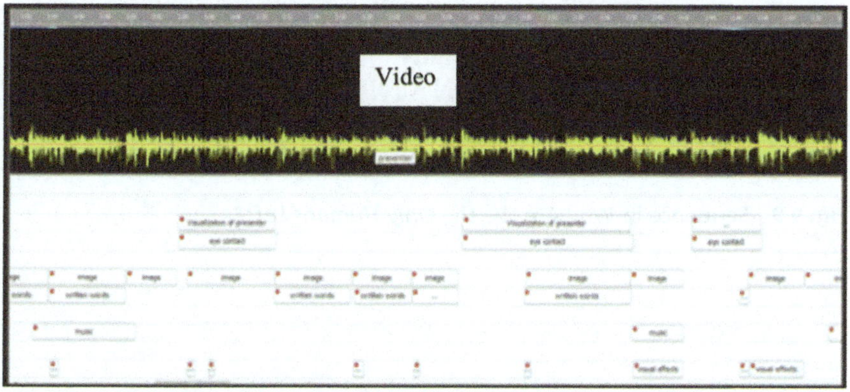

Fig. 9.7 Annotated sequence in *Megalodon*

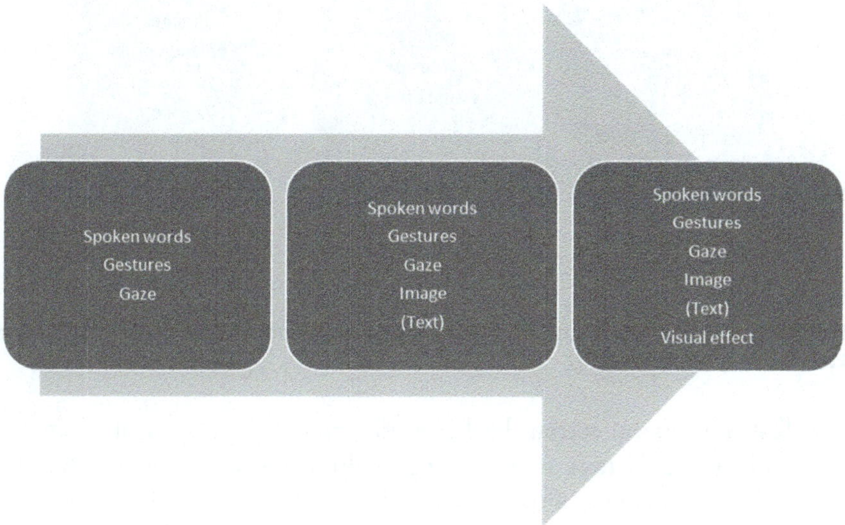

Fig. 9.8 Description of the increasing complexity of the multimodal ensembles

how a visualisation of the presenter speaking, gesturing and gazing at the camera is followed by an *image* accompanied by *written text*, which is then animated through *visual effect*. The cycle is then repeated.

The increasing complexity of these ensembles is represented in Fig. 9.8.

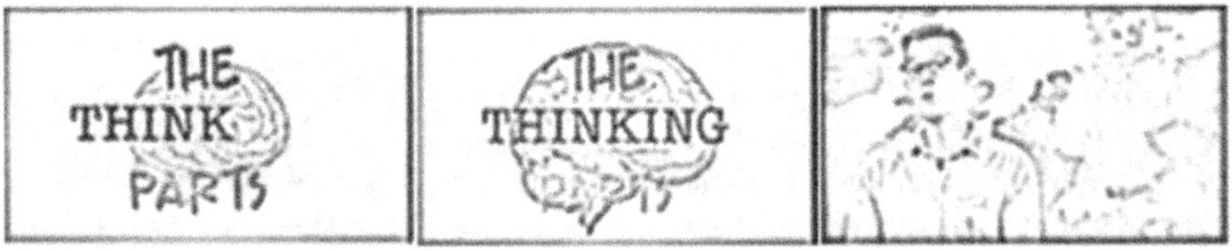

Fig. 9.9 Sequence in *Why Are We the Only Humans Left?*

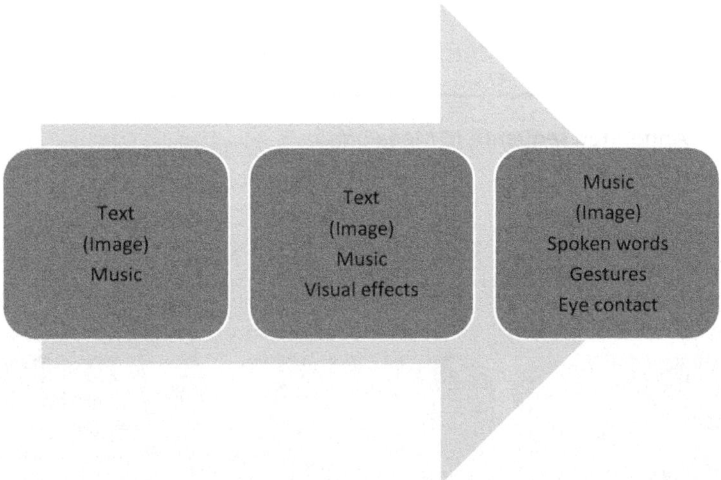

Fig. 9.10 Description of the evolution of the multimodal ensemble in *Why Are We the Only Humans Left?*

(2) Text appears on screen, background *music* is heard, visual *effects* are applied (e.g., letters move, change colour, resize), an *image* may be added and a fading of the screen leads to a visualisation of the presenter speaking. This sequence is used in *Why Are We the Only Humans Left?* to explicitly mark and name sections. An example can be seen in Fig. 9.9.

The evolution of the ensemble is represented in Fig. 9.10.

Our second research question (RQ2) probed into the recontextualisation strategies that are used in these videos. As explained in Sect. 2, the identification of strategies was informed by taxonomies suggested in previous literature (Luzón, 2019; Carter-Thomas & Rowley-Jolivet, 2020; Valeiras-Jurado et al., 2018), but as it was the case with modes, we adopted a corpus-driven, inductive approach, and contributed to these taxonomies when needed. The multimodal analysis revealed that the following recontextualisation strategies are present in our corpus (see Table 9.3).

Table 9.3 Recontextualisation strategies present in our corpus

Rhetorical aim	Strategy	Description
Strategies to tailor information	Selection of content	Excluding what is considered too difficult or not interesting
	Paraphrases	Repeating using different words
	Analogies	Comparisons between one thing and another for the purpose of clarification
	Examples from everyday life	Bringing content closer referring to common everyday experiences
	Scenarios, stories	That is, "picture yourself"
	Visual representations	Pictures, video, iconic gestures, etc. that clarify content (see Fig. 9.4)
Strategies to engage the audience	Emphasis	Highlighting what is considered surprising, novel or immediately applicable
	Attention retention	Making sure the viewer stays until the end of the video
	Dialogic involvement	Questions and addresses to the audience, inclusive pronouns
	Anticipation and control of responses	Presenting something as known, as interesting or relevant, anticipating audience questions or remarks
	Humour	Jokes, irony (Fig. 9.3)
Strategies to build credibility	Rapport	Inclusive language, smiles
	Evaluation	For example, evaluative adjectives and intensifying adverbs
	Personalisation	Self-mentions, personal information
	Mentions of network	Conferences and other scientific events, colleagues, subscribers, naming the channel
	Mentions of patrons and supporters	For example, PBS studios, companies (Fig. 9.3)
	Mentions of scientific sources	Publications, interviews with experts
	Street credibility	References to popular culture and knowledge (e.g., films, books, characters, etc.)
	Cross-referencing	For example, mentioning other episodes (see Fig. 9.10)

"That's an episode well worth checking out for a deeper dive"

Fig. 9.11 Cross-referencing strategy in *Muon*

We also noticed that *selection of content* did not allow a time-aligned, annotation-based analysis, as the rest of the strategies. Rather, presenters select content beforehand, as they prepare the script. For this reason, we consider this strategy as a first step towards recontextualisation and a prerequisite for other strategies to happen. In fact, in the case of the Muon video, we can observe how selection of content leads into *cross-referencing*, as the speaker refers the audience to other episodes for a deeper discussion of content that is not included in the current one. Figure 9.11 shows an example.

It is also important to note that the boundaries between strategies are not clear-cut. It is not always possible to determine with accuracy where one ends, and another begins. This is the case, for example, of the strategies *emphasis* and *attention getting*, which in some cases can only be differentiated taking into account intensity and duration (i.e., we consider attention getting as sustained in time and more intense than emphasis).

Our third research question (RQ3) concerns the multimodal realisation of these strategies. A first outcome of our analysis is that even strategies that are rooted in the tradition of linguistic-based discourse analysis and are therefore conceived as being realised through language, can also be conveyed through a variety of modes. For example, analogies can be conveyed through *images*, *mentioning patrons* can be enacted through

"Yes, this will be on the test and it's also important for the next thing I'm going to say

Fig. 9.12 Example of modal density and complexity of *Muon* video

clothes (see Fig. 9.3) and cross-references to other episodes sometimes take the form of a *deictic gesture* pointing at a pop-up banner with the title of the episode. Figure 9.12 provides a better insight on the modal density and complexity that these videos can achieve.

In this excerpt of the video *Why the Muon g-2 Results Are So Exciting!* we notice a variety of modes at play. While *visualising the presenter*, we also see an *equation* accompanied by *written text*. The speaker introduces a humorous pun by addressing the viewer saying: "Yes, this will be on the test." Synchronously, he *gazes* at the camera, makes a *warning sign* pointing at the audience and he raises his eyebrows. Simultaneously, we can also see the PBS logo in the top left corner of the screen and in the presenter's T-shirt, which also shows the name of the channel Spacetime. This multimodal ensemble realises a series of recontextualisation strategies. The presenter is introducing humour and building a fictional dialogue with the viewer. He is also building rapport and directing the response of the audience by showing this is relevant to understand what is coming next. Finally, his *clothing* provides mentions of patrons and supporters and mentions of network (PBS and Spacetime). Table 9.4 shows the recontextualisation strategies in this excerpt and their modal realisations.

A second example of the multimodal realisation of strategies is provided in Fig. 9.13, which corresponds to the beginning of the video *Why Are We the Only Humans Left?*

Look around and you'll meet some pretty smart animals, but there's no species quite like us.

Fig. 9.13 Multimodal realisation of recontextualisation strategies in *Why Are We the Only Humans Left?*

Table 9.4 Description of multimodal realisation of recontextualisation strategies in *Muon*

Engaging the audience			Building credibility		
Dialogic involvement	Humour	Anticipation and control of responses	Rapport	Mentions of patrons	Mentions of network
Spoken words Gestures Eye contact Face expression	Spoken words Gestures Eye contact Face expression	Spoken words Gestures Eye contact Face expression Equation Written text	Spoken words Gestures Eye contact Face expression	Clothing	Clothing

In this excerpt, there is once more a variety of modes contributing meaning. We can notice the use of *image* and *text*. The image is not static but moves through *visual effects*. At the same time, the presenter is *talking* and *gazing* at the camera while *gesturing* (e.g., pointing at himself as he says "us"). All throughout, *music* can be heard. This multimodal ensemble realises a number of recontextualisation strategies. On the one hand, the *music*, *image* and *visual effects* are a way of getting the attention of the audience at the onset of the video. The images also act as a visual representation of the comparison that the speaker is establishing between humans and animal species. At the same time, they also refer to popular culture (i.e., TV contests) to build street credibility, that is, a common framework based on shared cultural values rather than scientific know-how. Finally, the direct address to the audience ("look around") and the use of "us" as the speaker keeps *eye contact* and *points* inwards at himself are all a way to create dialogic involvement. Table 9.5 summarises the multimodal recontextualisation strategies in this excerpt.

The analysis also showed that some strategies tend to be realised through multimodal ensembles that show a preference for a particular mode. In other words, we can notice certain correlations between some strategies and some modal realisations. This is the case of the strategy visual representations that is always realised through an ensemble that contains *image*, *video* or *artefacts*, and may additionally contain other modes like *spoken words*, *gestures* or *music*. Figure 9.14 from *What Is the Magic Russian Diamond?* shows the use of a coin as a reference to explain the size of the diamond.

Table 9.5 Description of multimodal realisation of recontextualisation strategies in *Why Are We the Only Humans Left?*

Tailoring information	Engaging the audience		Building up credibility
Visual representation	Attention getting	Dialogic involvement	Street credibility
Image Written words Visual effects	Image Visual effects Music	Spoken words Gestures Music Eye contact	Images Written words Visual effects

Fig. 9.14 Visual representation in *What Is the Magic Russian Diamond?*

Fig. 9.15 Multimodal realisation of emphasis in *Why Are We the Only Humans Left?*

This is probably a result of the affordances provided by each mode. The strategy visual representations, as its name indicates, is highly dependent on visual elements as a medium (i.e., its material form is inherently visual). It is therefore not surprising that the modes that are used to enact this strategy are also based on visual elements as media: an *image*, a *video* (which can be considered a moving image) or a physical object that is manipulated for the viewer to see. These visual elements also have the ability to convey meaning in a more idiosyncratic way and therefore can facilitate and speed the clarification process, in the same way as a map indicates the way faster and more effectively than written or spoken directions.

We find a similar tendency with the strategy emphasis, which tends to be realised through ensembles that contain *sound* and *visual effects*. An example from *Why Are We the Only Humans Left?* is shown in Fig. 9.15. In this excerpt of the video, we see a cartoon-like depiction of a Neanderthal disappear with a hissing sound and clouds appearing in its place.

We believe this is once more due to the affordances of each mode. *Visual* and *sound effects* have a dynamising effect on the video and therefore are a particularly suitable way to highlight certain elements and make them stand out for the viewer.

Finally, the multimodal analysis also revealed that presenters do not only use a variety of modes to realise recontextualisation strategies. In doing so, they also interact with their online media. For example, they seem to be very aware that viewers access their videos through the PBS webpage or through YouTube, and that they can make comments, or

"we have you covered, episode list in the description"

Fig. 9.16 *Muong* speaker interacting with the online media

browse to watch other videos from the episode list. Not only that, they also seem to be very conscious of how this information shows up for the viewers, to the extent that they even make *deictic gestures* up or down depending on where this information appears on screen. Figure 9.16 shows how the presenter *points down* with both hands when referring to the episode list that shows up below the player window on the website.

4 Conclusions

In this chapter, we have undertaken a multimodal analysis of recontextualisation in four science dissemination videos. The analysis has revealed that these videos display a high number of modes and rich, seamlessly woven multimodal ensembles that realise a number of recontextualisation strategies.

Concerning our RQ1, our results show that embodied modes that inevitably accompany oral discourse (i.e., *gestures*, *face expression*, *head movements*) contribute meaning, for example, directing responses of the audience, or providing emphasis. Also modes that are not so evident as meaning-making resources play a crucial role in recontextualising scientific content. *Artefacts*, for example, can provide visual representations that facilitate understanding of difficult concepts. Likewise, *clothing* can serve the purpose of displaying the names of patrons or achieve a humorous effect. The analysis has also brought to the fore the use of *filmic modes* as a consequence of the online media, which are particularly prominent in this genre. Presenters achieve a livelier and more effective interaction with their audiences if they show awareness of the media through which these audiences access their videos and the way the information is organised on their screens (e.g., where the comment list is, or where the title of the episode appears). Despite these modes being used with great flexibility and creativity, it is not possible to identify recurrent multimodal ensembles in these videos. However, the analysis of the videos has revealed some modal co-occurrences, for instance *spoken words* + *gestures* + *eye contact*, and some preferences in the use of modes, for example the use of *animated images* or *animated text*. Furthermore, some regularity in the sequencing of modes has also been identified. In these cases, the modes are sequenced in a regular pattern during a part of the video making the multimodal ensemble more complex and dense as the sequence progresses (see Figs. 9.8 and 9.9).

As for RQ2, our analysis shows how presenters use a variety of recontextualisation strategies to tailor the information to the assumed knowledge of the target audience, engage the audience and build credibility. These objectives are clear from the onset if we pay attention to the titles of the videos. Three of them take the form of a "why" question and the fourth is phrased as a "what" question. These titles raise curiosity and prepare the audience to expect content that is not only accessible and interesting, but also engaging, thus creating a receptive frame of mind. Results show that in some cases, the boundaries between strategies are not clear-cut, and it is not always possible to determine with accuracy where one ends, and another begins. Our corpus-driven approach has allowed us to revise and complete previous taxonomies with strategies that have not been applied to online genres so far, such as "anticipation and control of responses." (see Tables9.5)

Regarding RQ3, as our analysis has shown, recontextualisation strategies, although rooted in linguistic tradition, are realised multimodally. On the one hand, one multimodal ensemble can realise a number of recontextualisation strategies (see Tables 9.4). On the other hand, some strategies tend to be realised through multimodal ensembles that show a preference for a particular mode. In other words, certain correlations between some strategies and some modal realisations have been found (e.g., *visual representations-images*).

The corpus-driven approach adopted in this study and the multimodal analysis of the corpus selected have helped us revise and complete existing taxonomies of both modes and strategies, contributing in this way to our knowledge of science dissemination, digital genres and multimodality.

The study presented in this chapter is limited in scope, since the corpus includes four videos from two disciplines. Further research with larger corpora would help to identify genre traits and to further explore recurrent ensembles and sequences of modes. This would expand our understanding of this emerging genre and of multimodal recontextualisation practices.

Acknowledgements This research has been supported by the *Ministry of Science, Innovation and Universities* [grant number PGC2018-094823-B-I00] and Research Promotion Plan at Universitat Jaume I (Spain) [grant number UJI-B2020-09].

References

Argyle, M., Ingham, R., Alkema, F., & McCallin, M. (1981). The different functions of gaze. In A. Kendon (Ed.), *Non-verbal communication, interaction and gesture* (pp. 221–237). Mouton.

Bezemer, J. A., & Kress, G. (2008). Writing in multimodal texts: A social semiotic account of designs for learning. *Written Communication, 25*(2), 166–195.

Boy, B., Bucher, H.-J., & Christ, K. (2020). Audiovisual science communication on TV and YouTube. How recipients understand and evaluate science videos. *Frontiers in Communication, 5*, 1–18.

Carter-Thomas, S., & Rowley-Jolivet, E. (2020). Three minute thesis presentations: Recontextualisation strategies in doctoral research. *Journal of English for Academic Purposes, 48*, 1–14.

Goldman, R., Erickson, F., Lemke, J., & Derry, S. J. (2007). Selection in video. In S. Derry (Ed.), *Guidelines for video research in education: Recommendations from an expert panel* (pp. 19–27). Data Research and Development Center at the University of Chicago.

Jewitt, C., & Kress, G. (Eds.). (2003). *Multimodal literacy*. Peter Lang.

Kastberg, P. (2011). Knowledge asymmetries: Beyond 'to have and have not'. *Fachsprache: Internationale Zeitschrift für Fachsprachenforschung -didaktik und Terminologie, 33*(3–4), 137–151.

Kendon, A. (2004). *Gesture: Visible action as utterance*. Cambridge University Press.

Kress, G. (2010). *Multimodality: A social semiotic approach to contemporary communication* (2nd ed.). Taylor & Francis.

Ladewig, S. H. (2013). Creating multimodal utterances: The linear integration of gestures into speech. In C. Muller, A. Cienki, E. Fricke, S. Ladewig, D. McNeill, & S. Teßendorf (Eds.), *Body-language-communication: An international handbook on multi-modality in human interaction* (Vol. I, pp. 1662–1677). De Gruyter Mouton.

Luzón, M. J. (2019). Bridging the gap between experts and publics: The role of multimodality in disseminating research in online videos. *Ibérica. Journal of the European Association of Languages for Specific Purposes, 37*, 167–192.

Luzón, M. J., & Pérez-Llantada, C. (2019). Connecting traditional and new genres: Trends and emerging themes. In M.-J. Luzón & C. Pérez-Llantada (Eds.), *Science communication on the Internet: Old genres meet new genres* (pp. 1–18). https://doi.org/10.1075/pbns.308.01luz

Luzón, M. J., & Pérez-Llantada, C. (2022). *Digital genres in academic knowledge production and communication. Perspectives and practices*. Multilingual Matters. https://doi.org/10.21832/9781788924726

Müller, C. (2013). Gesture as a medium of expression: The linguistic potential of gestures. In C. Muller, A. Cienki, E. Fricke, S. Ladewig, D. McNeill, & S. Teßendorf (Eds.), *Body-language-communication: An international handbook on multi-modality in human interaction* (Vol. I, pp. 202–217). De Gruyter Mouton.

O'Halloran, K. L., Tan, S., & Marissa, K. L. E. (2015). Multimodal analysis for critical thinking. *Learning, Media and Technology, 42*, 147–170.

OECD. (2015). *Making open science a reality, OECD science, technology and industry policy papers*. OECD Publishing (No. 25).

Paltridge, B. (2012). *Discourse analysis* (2nd ed.). Bloomsbury.

Pérez-Llantada, C. (2021). Genres and languages in science communication: The multiple dimensions of the science-policy interface. *Language & Communication, 78*, 65–76.

Querol-Julián, M. (2011). *Evaluation in discussion sessions of conference paper presentations: A multimodal approach*. LAP Lambert Academic Publishing GmbH & Co. K.G., Saarbrücken.

Rowley-Jolivet, E., & Carter-Thomas, S. (2019). Scholarly soundbites. Audiovisual innovations in digital science and their implications for genre evolution. In M.-J. Luzón & C. Pérez-Llantada (Eds.), *Science communication on the Internet: Old genres meet new genres* (pp. 81–106).

Rowley-Jolivet, E., & Carter-Thomas, S. (2020). Three minute thesis presentations: Recontextualisation strategies in doctoral research. *Journal of English for Academic Purposes, 48*, 1–14. https://doi.org/10.1016/j.jeap.2020.100897

Ruiz-Madrid, M. N. (2021). A multimodal discourse approach to research pitches. *Journal of English for Academic Purposes, 52*, 1–15.

Sancho Guinda, C. (Ed.). (2019). *Engagement in professional genres*. John Benjamins.

Scotto di Carlo, G. (2014). The role of proximity in online popularizations: The case of TED talks. *Discourse Studies, 16*(5), 591–606.

Scotto di Carlo, G. (2015). Stance in TED talks: Strategic use of subjective adjectives in online popularisation. *Ibérica: Journal of the European Association of Languages for Specific Purposes, 29*, 201–221.

Stöckl, H. (2014). Typography. In S. Norris & C. D. Maier (Eds.), *Interactions, images and texts. A reader in multimodality* (pp. 281–295). Walter de Gruyter.

Valeiras-Jurado, J. (2019). Modal coherence in specialised discourse: A case study of persuasive oral presentations in business and academia. *Ibérica: Revista de la Asociación Europea de Lenguas para Fines Específicos, 37*, 87–114.

Valeiras-Jurado, J., & Bernad-Mechó, E. (2022). Modal density and coherence in science dissemination: Orchestrating multimodal ensembles in online TED talks and Youtube science videos. *Journal of English for Academic Purposes, 58*.

Valeiras-Jurado, J., & Ruiz-Madrid, M. N. (2019). Multimodal enactment of characters in conference presentations. *Discourse Studies, 21*(5), 561–583.

Valeiras-Jurado, J., Ruiz-Madrid, N., & Jacobs, G. (2018). Revisiting persuasion in oral academic and professional genres: Towards a methodological framework for Multimodal Discourse Analysis of research dissemination talks. *Ibérica: Revista de la Asociación Europea de Lenguas para Fines Específicos, 35*, 93–118.

van Leeuwen, T. (2006). Towards a semiotics of typography. *Information Design Journal, 14*(2), 139–155.

Vicente-Sáez, R., & Martínez-Fuentes, C. (2018). Open science now: A systematic literature review for an integrated definition. *Journal of Business Research, 88*, 428–436.

Xia, S. A., & Hafner, C. A. (2021). Engaging the online audience in the digital era: A multimodal analysis of engagement strategies in TED talk videos. *Ibérica: Revista de la Asociación Europea de Lenguas para Fines Específicos, 42*, 33–58.

Zhang, Y., & O'Halloran, K. L. (2013). From popularization to marketization: The hypermodal nucleus in institutional science news. In E. Djonov & S. Zhao (Eds.), *Critical multimodal studies of popular discourse* (pp. 172–190). Routledge.

Part IV

Engaging the Audience Through Science Bites

10

Three-Minute Thesis Presentations: Engaging the Audience Through Multimodal Resources

Vicent Beltrán-Palanques

1 Introduction

Ongoing developments in digital communication along with the democratisation of science are certainly contributing to the evolution and emergence of academic genres (Luzón & Pérez-Llantada, 2019, 2022). Technology and digital resources bring innovation and change in genre construction (Lorés & Diani, 2021), offering researchers unique dissemination opportunities to share knowledge, popularise science and reach broader audiences. This ever-increasing scenario poses communication challenges for researchers in terms of the expression of ideational, textual and interpersonal meanings (Halliday, 1978). The present study focuses on Three-Minute Thesis (3MT) presentations, a relatively new genre that is becoming increasingly widespread in academia. This popularity may be related to the interest in making science available to a general audience and the drive to internationalise and maximise institutional visibility

V. Beltrán-Palanques (✉)
Universitat Jaume I, Castelló, Spain
e-mail: vbeltran@uji.es

R. Plo-Alastrué, I. Corona (eds.), *Digital Scientific Communication*,
https://doi.org/10.1007/978-3-031-38207-9_10

worldwide. A 3MT presentation can be considered a dissemination genre that conforms to the rising trend to share research with a general audience (Carter-Thomas & Rowley-Jolivet, 2020). Throughout this spoken genre, doctoral students explain their research in a maximum of 3 minutes. In 3MT competitions, judges focus on PhD students' communication skills and their ability to transmit content to non-expert audiences (Carter-Thomas & Rowley-Jolivet, 2020). The first 3MT competition took place at the University of Queensland in 2008. Since then, over 900 universities in more than 85 countries are celebrating 3MT competitions.[1]

3MT competitions are usually organised to be held in person and then digitised and hosted online. A key feature of 3MT presentations is their brevity, which seems to be common in other digital genres (e.g., scientific podcasts and YouTube videos) (Rowley-Jolivet & Carter-Thomas, 2019) or other spoken genres such as Elevator Pitch (Beltrán-Palanques, 2022) and PechaKucha presentations (Harinck & van Leeuwen, 2020; Querol-Julián & Beltrán-Palanques, 2021). Time constraints on 3MT presentations and their specific communicative aims may present challenges for doctoral students, who find themselves at the beginning of their academic careers using English generally as a lingua franca. 3MT presentations permit doctoral students to showcase their communication skills to succinctly explain their ongoing research engagingly to non-expert audiences.

In the last few years, research interest in this genre has increased. Hu and Liu (2018) conducted a pioneering study to examine both the rhetorical structure of 3MT presentations from varied fields of expertise and whether there was variation across disciplines. Results indicated that 3MT presentations were articulated through six obligatory moves (i.e., orientation, rationale, purpose, methods, implication and termination) and two optional moves (i.e., framework and results). Furthermore, results revealed that regardless of the homogeneity of the moves, differences were found in a few moves (i.e., framework, methods and results). These findings provided useful insights into the way 3MT presentations are textually organised. The study conducted by Rowley-Jolivet and

[1] https://threeminutethesis.uq.edu.au/about.

Carter-Thomas (2019) focused on four digital audiovisual soundbites, namely, podcasts of popular science and a research journal website, authors' videos and 3MT presentations. The authors analysed their content, rhetorical structure and communicative strategies in relation to research articles and doctoral dissertations. As for 3MT, the authors argue that, in addition to representing a standardised presentation format, it is characterised by the simplification of the scientific content. Furthermore, through 3MT presentations, presenters attempt to engage their audience and narrate a story. The focus, therefore, seems to be on how content is conveyed. In this sense, the communication skills that doctoral students must develop to deliver a 3MT presentation seem to be far from those expected in other genres, such as a PhD defence. Carter-Thomas and Rowley-Jolivet (2020) also explored the development of 3MT presentations and the varied recontextualising strategies presenters use to share knowledge and engage the audience. Furthermore, drawing on Luzón (2013), the authors identified a series of strategies presenters used to establish interpersonal meaning, including titles, visual impact (e.g., striking visual images on the slide, physical presence), personalisation to create proximity (e.g., use of the first-person pronoun, stories), questions, and humour and street cred.

Taking a corpus-based approach and following Hyland (2005), Hyland and Zou (2021) examined how 3MT presenters articulated their positions and showed stance towards their research and audiences. Results highlighted the relevance of upholding stance across disciplines. In the case of hard sciences, presenters seemed to project stance by asserting certainty in the reliability of the information, whereas in social sciences, presenters appeared to share a more authorial self. Drawing on Hyland (2005), Qui and Jiang (2021) analysed 3TM presentations from several disciplines to identify the interactional features presenters used to take stance and enact engagement. Results disclosed that through 3MT, presenters used first-/second-person pronouns to address their immediate audience. Moreover, as the authors indicated, 3MT presentations differ from other academic spoken genres, mainly due to the heterogeneous target audience.

These studies not only serve to better learn about this new genre but also to understand that presenters face the need to develop specific communicative skills to construct meaning effectively. For students to succeed in 3MT presentations, they should learn how to express ideational, textual and, most importantly, interpersonal meanings. The findings from these studies invite reflection on how to approach 3MT presentations in EAP courses and workshops. For example, the study presented by Hu and Liu (2018) describes the rhetorical structure of 3MT presentations. The results of the study by Rowley-Jolivet and Carter-Thomas (2019) can contribute to the understanding of how to express content to reach wider audiences and show proximity. The studies conducted by Carter-Thomas and Rowley-Jolivet (2020), Hyland and Zou (2021) and Qui and Jiang (2021) appear to offer insights into how interpersonal meaning is constructed and established with the audience.

So far, studies on 3MT presentations have focused principally on the linguistic nature of 3MT presentations, except for Carter-Thomas and Rowley-Jolivet (2020) who briefly report on embodied actions. Nevertheless, research on pedagogy and the contribution of multimodality is unexplored. Delivering a 3MT presentation can be quite complex (e.g., time constraints, content expression, or textual organisation). Therefore, it would be advisable for EAP teachers to equip students with the specific communicative skills they may need to construct meanings through this genre (Hu & Liu, 2018; Hyland & Zou, 2021). Doctoral students should learn to synthesise, simplify and adapt complex disciplinary content to non-expert audiences. Moreover, due to the conversational and audience-oriented styles, doctoral students should gain an understanding of how to establish interpersonal meaning with the audience. Somehow, throughout 3MT presentations, doctoral students have the opportunity to talk about their ongoing research in the form of an engaging story (Rowley-Jolivet & Carter-Thomas, 2019). Engagement is essential in 3MT presentations, as well as a key judging criterion. Thus, engagement should be a core aspect to be addressed in any training course/workshop. In addition, multimodality should be introduced to make students aware of how to use and combine non-verbal resources to co-express and support meaning construction.

The present study adopts both a pedagogical and analytical perspective. Pedagogically, we discuss an already-implemented multimodal genre-based approach to teaching 3MT presentations to doctoral students. Then, we explore how two doctoral students organise discourse and express interpersonal meaning, as well as how the latter is conveyed multimodally through 3MT presentations.

2 Pedagogical Approach

The genre of 3MT presentations requires special attention from a pedagogical viewpoint. This is largely due to the characteristics of the genre itself and the skills presenters should develop to construct meaning effectively. Notwithstanding the growing popularity of 3MT presentations in academia, no research has yet addressed this issue. In this section, we briefly present a pedagogical approach adopted for the teaching of 3MT presentations. This pedagogical model has been implemented within an EAP workshop purposely designed to train doctoral students at Universitat Jaume I (Spain) (Beltrán-Palanques, 2023). The genre of 3MT presentations started to be implemented as part of the workshop during the 2020 academic course.

Specifically, we drew on previous multimodal genre-based approaches to train EAP/EAP students (Querol-Julián & Fortanet-Gómez, 2019; Querol-Julián & Beltrán-Palanques, 2021). This pedagogical approach allows students to become familiar with the conventions of 3MT presentations through sample exploration and construction. The teaching/learning cycle is composed of three phases, namely joint deconstruction, joint construction and independent construction (Martin & Rose, 2005). In the first two phases, the lecturer provides direct support, while in the third phase, the students become more autonomous. The deconstruction and joint construction phases are completed during the class session, to which approximately 90 minutes are devoted, and the independent construction is carried out autonomously.

During the deconstruction phase, the lecturer presents and defines the genre, paying particular attention to its communicative aim (Carter-Thomas & Rowley-Jolivet, 2020) and rhetorical moves (Hu & Liu, 2018). The joint construction phase is devoted to discussing the guidelines for the presentation, which can be found on the official website of the Three-Minute Thesis (3MT®) competition website (The University of

Queensland[2]). To further support students, the lecturer also shares with them additional resources (e.g., University of Surrey[3]). In addition, the lecturer provides explicit information concerning the expression of interpersonal meaning and the deployment of non-verbal resources (e.g., looking at the webcam, maintaining a frontal position and making use of hand and arm gestures). Finally, in the independent construction phase, students prepare and rehearse their presentations autonomously. Then, they video-record their 3MT presentation and submit it. Allowing students to prepare an asynchronous presentation can be beneficial as they have opportunities to rehearse and edit the digital document.

The purpose of the pedagogical proposal is to familiarise doctoral students with this digital genre in which researchers endeavour to convey and broaden scientific knowledge while involving the audience and promoting their critical thinking. Through this pedagogical model, students can reflect on how to talk about their ongoing research in a simplified and engaging way, while paying attention to the variety of communicative modes available to them for constructing meaning.

3 Methodology

3.1 Description of the Context and Participants

This study draws on an EAP workshop intended for doctoral students held at Universitat Jaume I in Spain. As part of this workshop, one of the final projects that doctoral students are required to prepare is a 3MT presentation. Considering that the spoken genre of 3MT presentations started to be implemented during the 2020 academic course, currently, over 50 students have participated in the course. The corpus consists of 54 presentations of doctoral students from diverse disciplines: Health Science (HS) (n=6), Humanities and Arts (HA) (n=13), Social Science (SS) (n=18) and Technical Science (TS) (n=17). The total length of the

[2] https://threeminutethesis.uq.edu.au/resources/3mt-competitor-guide.

[3] https://www.surrey.ac.uk/researcher-development-programme/researcher-community/three-minute-thesis.

recordings is 10,135 seconds (2 hours 48 minutes 55 seconds, mean: 187.68 seconds). Students were expected to produce a three-minute presentation (180 seconds); however, some of them did not meet this requirement. Nonetheless, longer or shorter videos were not discarded from the corpus. This is because the data consist of a compilation of students' presentations delivered as part of their training in the EAP workshop.

The dataset used in the current study stems from the 3MT presentations corpus above presented. For the present study, two 3MT presentations were chosen. The rationale for the election of the two participants was as follows: each participant belonged to a specific field of specialisation (i.e., SS and HA) and their 3MT presentations received a good final grade. As for the assessment, the presentations were rated considering the presenters' delivery of content, use of engagement strategies, language proficiency and adherence to the format.

The first participant was in the field of Education and her proficiency level was B2. On the other hand, the second participant belonged to the area of Applied Linguistics and his proficiency level was C2.

3.2 Data Analysis and Procedure

The study consists of two parts. The first part is the analysis of the moves followed to construct the 3MT presentation and the linguistic interpersonal strategies the presenters used to engage the audience. The second part centres on how presenters instantiated those interpersonal strategies using varied semiotic resources.

For the first analysis, the two 3MT presentations were carefully watched to identify the varied rhetorical steps used by the participants to construct the speech. Specifically, we followed the classification of moves found by Hu and Liu (2018) (i.e., orientation, rationale, purpose, framework, methods, results, implication and termination). The following step was to explore the interpersonal linguistic strategies the presenters employed to connect with the audience. We inductively developed a coding scheme of interpersonal strategies as they emerged from the dataset. For the second analysis, we drew on the interpersonal strategies identified in the first analysis to determine how they were constructed multimodally (Morell et al., 2020, 2022).

The data analysis was supported by the multimodal annotation software ELAN (Wittenburg et al., 2006). The software permitted the annotation of data for each analysis. As for the first analysis, the software was run to annotate moves and interpersonal strategies. For the second analysis, the software made it possible to annotate a series of non-verbal communicative modes that the presenters deployed when enacting interpersonal strategies.

In line with previous studies (e.g., Beltrán-Palanques & Querol-Julián, 2018; Beltrán-Palanques, 2019; Morell et al., 2022), the following modes were taken into account: speech, gestures (McNeill, 1992), facial expressions, gaze and posture. An ELAN template containing 7 tiers and specific controlled vocabulary was configured (Table 10.1). To ensure interrater reliability, the results of the analysis were revised and discussed with an external researcher.

Table 10.1 Controlled vocabulary

Tier	Controlled vocabulary
Rhetorical step	Orientation
	Rationale
	Framework
	Purpose
	Methods
	Implication
	Termination
Interpersonal meaning	Referential "you"
	Question
	Exclusive "we"
	Inclusive "we"
	Self-reference "I"
Gesture	Beat
	Deictic
	Iconic
	Metaphoric
	Pointing
	Rest position

(*continued*)

Table 10.1 (continued)

Face	Frowning Neutral Raising eyebrows Smile
Gaze	Frontal Left Right Up Down
Posture	Frontal (webcam) Lateral

In addition to this, for both analyses, we also obtained the annotation density plot to visualise the distribution of annotations (both verbal and non-verbal) over each of the presentations. This was done using the ELAN software.

4 Results and Discussion

4.1 Move Organisation and Enactment of Interpersonal Meaning

Having carefully analysed each of the videos, it appeared that the two presenters followed the obligatory rhetorical moves found in Hu and Liu (2018) and made use of other optional steps. Specifically, the two presenters organised their discourse according to the orientation, rationale, purpose, methods, implication and termination moves. Presenter 1 also covered the framework and results move, while Presenter 2 the framework move. Table 10.2 shows the length and duration of the moves of each presentation.

The length of each of the moves was different, with the orientation (Presenter 1) and rationale (Presenter 2) moves being longer than the rest and the termination move being the shortest in both cases. The length of the orientation and rationale move for each presenter was especially striking when comparing them. Presenter 1 elaborated on a

Table 10.2 Length of the moves

Move	Begin time	End time	Duration
Presenter 1			
Orientation	00:00:03.919	00:01:07.800	00:01:03.881
Rationale	00:01:07.839	00:01:30.967	00:00:23.128
Purpose	00:01:30.967	00:01:54.322	00:00:23.355
Framework	00:01:54.330	00:02:11.124	00:00:16.794
Methods	00:02:11.128	00:02:36.876	00:00:25.748
Results	00:02:36.914	00:02:48.462	00:00:25.748
Implication	00:02:48.500	00:03:07.307	00:00:18.807
Termination	00:03:07.364	00:03:09.557	00:00:02.193
Presenter 2			
Orientation	00:00:01.104	00:00:32.800	00:00:31.417
Rationale	00:00:34.086	00:01:42.527	00:01:08.441
Framework	00:01:42.527	00:02:12.813	00:00:30.286
Purpose	00:02:13.155	00:02:22.522	00:00:09.367
Methods	00:02:22.680	00:02:43.533	00:00:20.853
Implication	00:02:43.744	00:02:55.449	00:00:11.705
Termination	00:02:55.633	00:02:58.387	00:00:02.754

long orientation move (00:01:03.881) and a short rationale move (00:00:23.128). Presenter 2 acted oppositely, producing a short orientation move (00:00:31.417) and a long rationale move (00:01:08.441). Therefore, it seems that Presenter 1 made a special effort to connect with the audience from the beginning while little time was devoted to stating the rationale of the study. Conversely, Presenter 2 dedicated more time to the explanation of the rationale of the study but not so much to the orientation. Regardless of the length, throughout these two rhetorical steps, the two presenters shared some research content and established the motivation for the study, while trying to engage the audience (Hu & Liu, 2018).

One of the main limitations of this presentation format is time (Carter-Thomas & Rowley-Jolivet, 2020). However, despite this limitation, the two presenters managed to organise the speech, transmit the content and engage their audiences. Following the 3MT conventions, the content was conveyed in a simplified manner in the form of a story and using one single static visual.

At the beginning of the oration, the two presenters introduced the topic and provided some background information by calling the audience's attention (orientation move). Then, the presenters justified the need for their research (rationale move). Presenter 2 also provided some theoretical information to support the basis of his study (framework move) right after the rationale move. Having involved the audience in the talk and briefly described the rationale and theoretical background of the study, the two presenters stated the purpose (purpose move) and the methodology of their research (methods move). Presenter 1 also commented on some possible findings (result move). Then, the two presenters briefly referred to the contribution of their study and why it was relevant (implication move). Finally, the two presenters thanked the audience (termination move).

As for the expression of interpersonal meaning, we explored the typology and frequency of strategies the two presenters used throughout their 3MT presentations. The analysis was carried out using the ELAN multiple-layer search engine, which showed a total of 60 hits in 60 annotations (out of 421 annotations) in the two 3MT presentations (i.e., Presenter 1, n=32; Presenter 2, n=28). Table 10.3 provides a summary of the typology and frequency of interpersonal strategies identified in the two 3MT presentations.

As shown in Table 10.3, six interpersonal strategy types were found in the dataset. The most frequent strategy was referential "you" (34), followed by exclusive "we" (9) and self-reference "I" (9). Considering these results, we proceeded to examine the distribution of the interpersonal strategies within the varying moves. To best illustrate the distribution, we obtained the annotation density plot from the ELAN (Figs. 10.1 and 10.2).

Table 10.3 Typology and frequency of interpersonal strategies found in the dataset

Interpersonal strategy	Frequency	Percentage (%)
Referential "you"	34	56.67
Exclusive "we"	9	15.00
Self-reference "I"	9	15.00
Question	5	8.33
Inclusive "we"	3	5.00
Total	60	100

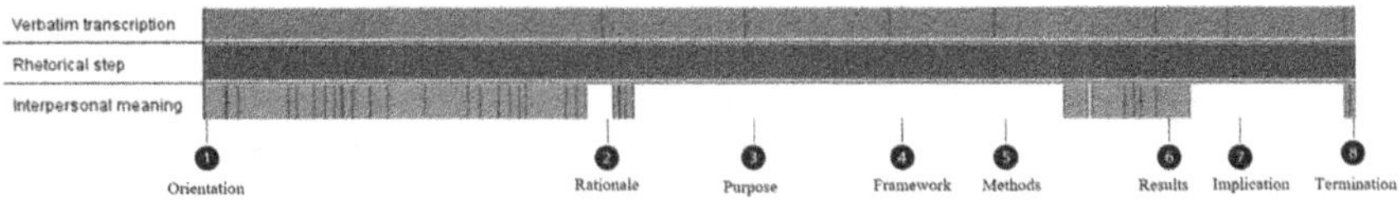

Fig. 10.1 Annotation density plot (ELAN). Presenter 1

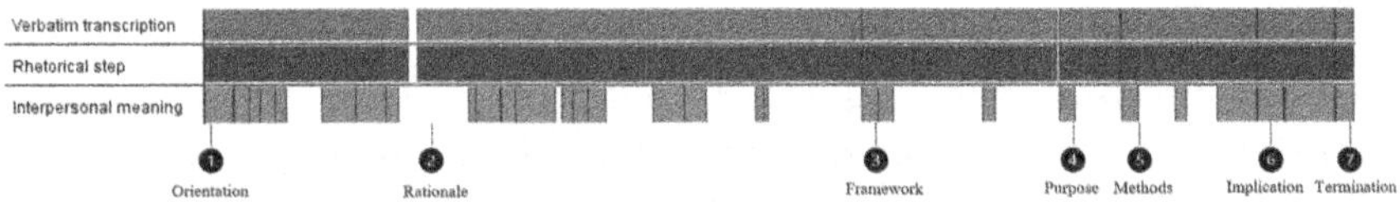

Fig. 10.2 Annotation density plot (ELAN). Presenter 2

Through the annotation density plot, we visualise the distribution of the interpersonal meaning within each of the moves. As observed, instances of interpersonal meaning predominated in the orientation and rationale moves. These results were expected since 3MT presenters at the beginning of the oration tend to engage the audience in interactive discourse (Hu & Liu, 2018; Carter-Thomas & Rowley-Jolivet, 2020). Instances were also found in other moves, such as in the methods and termination moves. For example, the presenters used interpersonal strategies in the method move to refer to their research. Likewise, in the termination move, the presenters appeared to use interpersonal strategies to probably conclude the speech (Carter-Thomas & Rowley-Jolivet, 2020). This result also suggests that the two presenters were aware of the relevance of establishing and holding the audience's attention throughout the entire 3MT presentation, especially at the start and end of the oration. The annotation density plot also shows that on some occasions, interpersonal strategies were not used. For example, Presenter 1 did not make use of interpersonal strategies in the purpose, framework, methods and implication move. The lack or reduced number of interpersonal strategies in certain rhetorical steps may be related to the specific communicative purpose that each one of them serves for the overall presentation.

This analysis should be complemented by the exploration of the typology and frequency of interpersonal strategies found throughout the varying moves in the two presentations. For this purpose, we also used the

Table 10.4 Distribution of interpersonal strategies in the moves

Moves	Strategies	Frequency	Percentage (%)
Orientation	Referential "you"	26	43.33
	Question	2	3.33
	Self-reference "I"	1	1.67
Sub-total 29			
Rationale	Referential "you"	5	8.33
	Inclusive "we"	3	5.00
	Self-reference "I"	2	3.33
	Exclusive "we"	2	3.33
	Question	1	1.67
Sub-total 13			
Purpose	Self-reference "I"	1	1.67
Sub-total 1			
Framework	Question	2	3.33
	Exclusive "we"	1	1.67
Sub-total 3			
Methods	Self-reference "I"	4	6.67
	Exclusive "we"	4	6.67
Sub-total 8			
Results	Self-reference "I"	1	1.67
Sub-total 1			
Implication	Exclusive "we"	2	3.33
Sub-total 2			
Termination	Referential "you"	3	5.00
Sub-total 3			
Total amount of strategies		60	100

ELAN multiple-layer search engine, which revealed a total number of 60 hits in 60 annotations (out of 421 annotations) in the two 3MT presentations. Table 10.4 shows a summary of the results.

As shown in Table 10.4, most interpersonal strategies were found in the orientation move (n=29), followed by the rationale (n=13) and methods (n=8) moves. The presenters appeared to principally use referential "you" in the orientation (n=26) move and occasionally in the rationale (n=5) and termination (n=3) moves. The choice of the engagement marker "you" may correspond to the presenters' attempt to directly address, engage and establish a sort of (emotional) proximity with the audience (Hyland, 2010; Luzón, 2013; Carter-Thomas & Rowley-Jolivet, 2020; Qui & Jiang, 2021). This is best illustrated in Examples 10.1 and

10.2, in which the presenter attempts to raise the audience's empathy and understanding of the situation she is describing. Specifically, the presenter seems to appeal to the social and emotional support of the audience. Through the realisation of these interpersonal resources, the presenter strives to engage the audience from the beginning of the presentation.

Example 10.1

Imagine that *you* are an eight-year-old child in a school, imagine that *you* are in a classroom with your peers, interacting with them, playing with them [...]. (Orientation move, Presenter 1)

Example 10.2

It must determine *your* childhood, *your* well-being [...]. (Rationale move, Presenter 1)

In Example 10.3 the presenter demonstrates gratitude and appreciation. It shows the use of "you" within the speech act of thanking to signal closing and politeness.

Example 10.3

Thank *you* for *your* attention. (Termination move, Presenter 1)

Results also reveal the use of exclusive and inclusive "we". The former was observed in the rationale (n=2), methods (n=4), implication (n=2) and framework (n=1) move, and the latter in the rationale move (n=3). The exclusive "we" relates to situations in which the presenters referred to themselves as researchers (Example 10.4), while the use of inclusive "we" allowed them to involve the audience (Example 10.5).

Example 10.4

[...] *we* want to describe a particular context [...]. (Methods move, Presenter 1)

Example 10.5
We use more than one semiotic mode to communicate [...]. (Implication move, Presenter 2)

As shown in these examples, the pronoun "we" is used for different purposes (Hardwood, 2005). Example 10.4 represents a situation in which the presenter referred to her professional/research sphere, thereby excluding the audience. On the other hand, in Example 10.5, the presenter used the inclusive "we" to involve the audience and build rapport.

Moreover, instances of self-reference with "I" were observed in the methods (n=4), rationale (n=2), orientation (n=1), purpose (n=1) and results (n=1) moves. The self-reference used by the presenters appeared to contribute to enhancing their role as researchers; yet as approachable people (Carter-Thomas & Rowley-Jolivet, 2020) trying to also connect with the audience (Examples 10.6 and 10.7).

Example 10.6
I know what you're thinking [...]. (Rationale move, Presenter 2)

Example 10.7
One of *my* (XXX) is the diary, where *I* write down everything *I* observe [...]. (Methods move, Presenter 1)

In Example 10.6 the presenter seems to use "I" to show empathy and mutual understanding while engaging and involving the audience. Example 10.7 differs from the previous example, as it appears that the presenter is trying to transmit an individualistic authorial presence (Hyland & Zou, 2021). In addition, the fact that 3MT presenters are expected to be more approachable and interactive may also favour the use of self-references to, for example, show stance and engagement (Hyland, 2006; Hyland & Jiang, 2017).

The use of questions was observed in the orientation (n=2), framework (n=2) and rationale (n=1) moves (Examples 10.8 and 10.9), probably to call the audience's attention indirectly and engage them (Luzón, 2013; Carter-Thomas & Rowley-Jolivet, 2020).

Example 10.8
How do you feel? (Orientation move, Presenter 1)

Example 10.9
But *what about lecturers?* (Framework move, Presenter 2)

In these two examples, the presenters seemed to be addressing the audience to raise empathy and make them reflect on specific situations. Therefore, the use of questions permitted the two presenters to promote the audience's involvement in the talk.

This first analysis allowed us to identify the way the two presenters organised their 3MT presentations in terms of moves and the use of interpersonal strategies to engage the audience. Results show that, in general, the training the participants received was effective in the sense that they followed the obligatory rhetorical steps found in Hu and Liu (2018), and even used optional ones. As for interpersonal strategies, overall, these results indicate that the two participants tended to exploit specific resources to engage the audience (e.g., referential "you", inclusive and exclusive "we"). The use of interpersonal resources was mainly found in the orientation and rationale moves, probably attributed to their inherently audience-oriented nature. However, although presenters predominantly used interpersonal resources in these two moves, they attempted to engage the audience constantly throughout the presentation. Overall, these results are in line with previous studies on 3MT presentations (e.g., Carter-Thomas & Rowley-Jolivet, 2020; Qui & Jiang, 2021) in which the use of pronouns such as "you" and "we" was proved to be recurrent. Indeed, the conversational, interactive and audience-oriented nature of 3MT presentations allows presenters to exploit certain pronouns that serve to directly address, connect and establish proximity with the audience (Qui & Jiang, 2021).

4.2 Multimodal Construction of Interpersonal Meaning

The second analysis in this study centred on the exploration of interpersonal meaning from a multimodal perspective. The point of departure for this second analysis was the results obtained in the first one regarding the moves and the expression of interpersonal meaning. For the second analysis, we first obtained the annotation density plot and then identified the recurrent multimodal ensembles. The ELAN software supported the examination of the data.

The annotation density plot for each presenter consisted of the following semiotic modes, that is, speech (rhetorical steps and interpersonal meaning), facial expressions, gaze, gestures and posture (Figs. 10.3 and 10.4).

Figures 10.3 and 10.4 visually represent the variety and distribution of multimodal resources the presenters used while enhancing interpersonal relationships with the audience. The results of the annotation density plot depicted here are in line with those found in the linguistic analysis,

Fig. 10.3 Annotation density plot (ELAN). Presenter 1

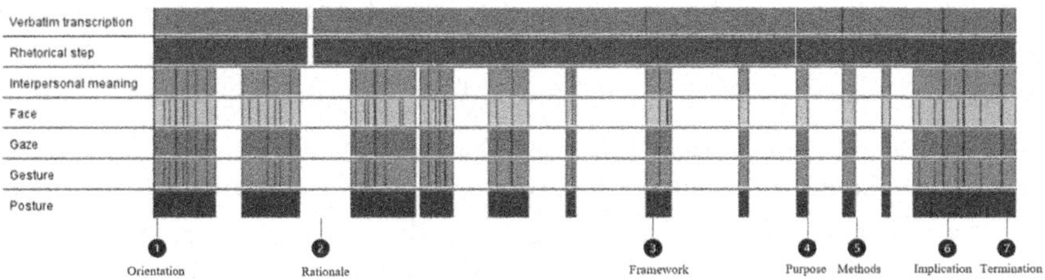

Fig. 10.4 Annotation density plot (ELAN). Presenter 2

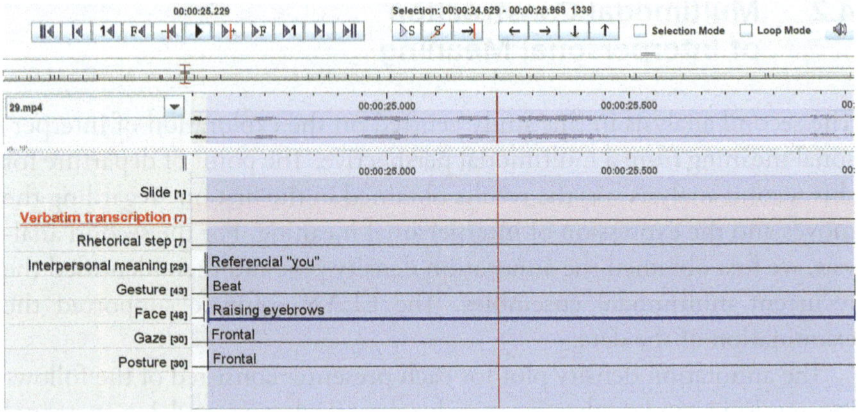

Fig. 10.5 An example of the annotation of a multimodal ensemble (Presenter 2, Orientation move)

which revealed the distribution of the presenters' expression of interpersonal meaning across the moves. The difference here lies in the inclusion of other semiotic modes, that is, facial expression, gaze, gestures and posture. The annotation density plot shown here offers an overview of the presenters' orchestration of modes to enhance interpersonal meaning throughout the 3MT presentations. Yet, these results should be viewed in relation to whether the communicative modes were used consistently through multimodal ensembles. By multimodal ensemble, we refer to the representation or orchestration of more than one communicative mode that contributes to the representation of meaning (Fig. 10.5).

As exemplified in Fig. 10.5, the orchestration of communicative modes forming a specific multimodal ensemble consists of the following: orientation move, interpersonal strategy-referential "you", facial expression-raising eyebrows, gaze direction-frontal, gesture-beat and posture-frontal.

To determine the multimodal ensembles, we calculated the variety of semiotic resources using the ELAN multiple-layer search engine. Specifically, speech, facial expressions, gaze direction and gestures were selected for this calculation. Posture was discarded since it was always frontal except for one situation, that is, Presenter 1 turning back (left gaze direction, lateral body posture and deictic gesture). The ELAN

Table 10.5 Recurrent multimodal ensembles as instantiated in the rhetorical moves

Multimodal ensemble	Frequency	Percentage (%)
Orientation, Referential "you", Neutral, Frontal, Iconic	18	10.53
Orientation, Referential "you", Raising eyebrows, Frontal, Iconic	14	8.19
Orientation, Referential "you", Raising eyebrows, Frontal, Beat	11	6.43
Orientation, Referential "you", Neutral, Frontal, Beat	9	5.26
Orientation, Referential "you", Raising eyebrows, Frontal, Deictic	7	4.09
Orientation, Referential "you", Neutral, Frontal, Deictic	6	3.51
Methods, Exclusive "we", Raising eyebrows, Frontal, Iconic	4	2.34
Methods, Exclusive "we", Neutral, Frontal, Iconic	4	2.34
Methods, Self-reference "I", Neutral, Frontal, Iconic	4	2.34
Orientation, Referential "you", Frowning, Frontal, Iconic	4	2.34
Rationale, Inclusive "we", Neutral, Frontal, Beat	4	2.34
Rationale, Inclusive "we", Neutral, Frontal, Iconic	4	2.34
Rationale, Referential "you", Neutral, Frontal, Iconic	4	2.34

multiple-layer search engine showed a total number of 171 hits in 171 annotations (out of 421 annotations) in the two videos, with 67 different multimodal ensembles. Table 10.5 shows the most recurrent multimodal ensembles. The data analysis revealed that some multimodal ensembles were more recurrent than others. For example, the following ensemble *orientation, referential "you", neutral, frontal, iconic* occurred 18 times, whereas other ensembles appeared only one time, for example, framework, exclusive "we", neutral, frontal, beat. A cut-off point was established; specifically, only the ensembles that were found to be repeated at least four times were considered. This was done to ensure representativeness within the study sample.

These results reveal that the presenters tended to enact interpersonal meaning through a broad range of multimodal ensembles. The multimodal combination of modes serves to make discourse more comprehensible (Campoy-Cubillo & Querol-Julián, 2015; Norte, 2018; Morell et al., 2022) for audiences, in general, but especially for non-expert audiences, as is the case of 3MT presentations. In addition, by orchestrating varied semiotic resources, 3MT presenters can best engage and involve their audiences in the talk. As results show, this may be true at least for the expression of referential "you", exclusive "we", self-reference "I" and inclusive "we". The most common multimodal ensembles were observed in the orientation move, followed by the methods and rationale moves. This result may be expected since, as seen in the previous section, the orientation (n=29), rationale (n=13) and methods (n=8) moves accounted for the largest number of interpersonal strategies.

In the orientation move (Figs. 10.6 and 10.7), the use of referential "you" came to the fore as an engagement marker. As for the rationale move (Figs. 10.8 and 10.9), multimodal combinations were found to be constructed through inclusive "we" and referential "you". Concerning the methods move (Figs. 10.10 and 10.11), results point to multimodal combinations with exclusive "we" and self-reference "I".

The referential "you" was widely used in these two presentations. Through the referential "you", presenters can directly address the audience and involve them in the speech (Hyland, 2002, 2010; Luzón, 2013; Rowley-Jolivet & Carter-Thomas, 2019). Self-reference "I" and exclusive and inclusive "we" were also found. While presenters can reinforce their authorial voice using self-reference "I" voice (Hyland & Zou, 2021), the use of "we" may allow them to either strengthen their position and

"Imagine that *you* are an eight-year-old child in a school"

Fig. 10.6 Orientation move, referential "you", facial expression-raising eyebrows, gaze direction-frontal, gesture-beat, posture-frontal

"Imagine *you* are in a classroom with *your* peers"

Fig. 10.7 Orientation move, referential "you", facial expression-raising eyebrows, gaze direction-upward and frontal, gesture-beat and iconic, posture-frontal

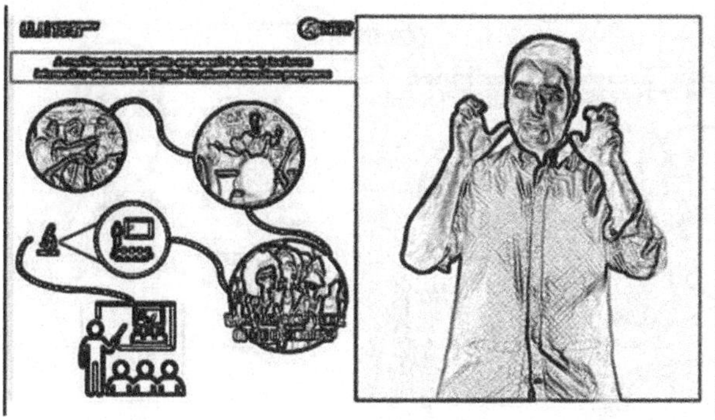

"when we speak"

Fig. 10.8 Rationale move, inclusive "we", facial expression-neutral, gaze direction-neutral, gesture-iconic, posture-frontal

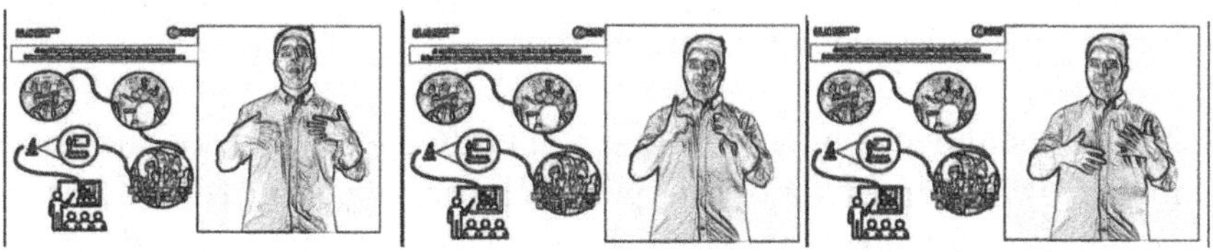

"*we* use more than one semiotic mode to communicate"

Fig. 10.9 Rationale move, inclusive "we", facial expression-neutral, neutral and frowning, gaze direction-neutral, gesture-beat, iconic and beat, posture-frontal

"*I* write down everything I observe, everything *I* experience and feel"

Fig. 10.10 Methods move, self-reference "I", facial expression-neutral, gaze direction-frontal, gesture-iconic, posture-frontal

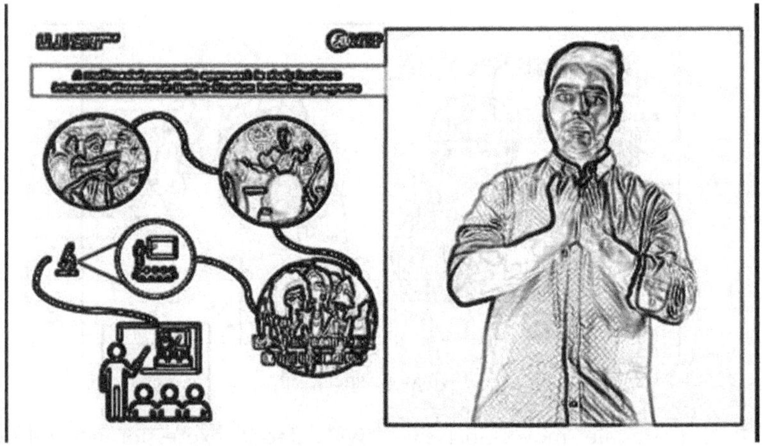

"in order to do this, we'll compile a corpus"

Fig. 10.11 Methods move, exclusive "we", facial expression-neutral, gaze direction-frontal, gesture-iconic, posture-frontal

exclude the audience or involve them (Hardwood, 2005). As shown in Table 10.5, the expression of interpersonal meaning occurred within specific multimodal ensembles, which together contribute to effectively engaging the audience. Results reveal the recurrent interplay of facial expressions, including neutrality, raised eyebrows, and frowning, which were used, for example, to express feelings and attitudes in relation to the spoken content. The two presenters tended to maintain a frontal gaze direction and posture. Embodied actions such as gaze and posture seemed to be characterised by the digital nature of the presentation, along with the lack of an on-site audience. Thus, it may be suggested that the presenters were focused on the webcam to directly address the audience. Moreover, results also show that the participants made use of a variety of gestures including beats, deictic and iconic (McNeill, 1992). In general, the use of gestures was found to make the discourse more dynamic, facilitate comprehension, reinforce the message, and directly address the audience. For instance, beat gestures helped to coordinate or emphasise parts of the speech and iconic gestures served to represent specific meanings. On occasion, deictic gestures were expressed with referential "you" as an ensemble to construct an engagement marker.

In short, the use and combination of varying semiotic resources allowed the two presenters to fulfil their communicative purposes and engage the audience throughout the 3MP presentation. Broadly speaking, the interaction of verbal and non-verbal resources permitted the two presenters to express their thoughts and feelings, show proximity and engage their audiences.

Based on the results obtained, it is suggested that the expression of interpersonal meaning was realised multimodally. Both participants were aware of the need to use and combine specific semiotic resources to engage their audiences. In this sense, the semiotic awareness of the presenters regarding the orchestration of modes was essential for them to construct interpersonal meaning effectively. It seems that the training the students received was useful since it contributed to their multimodal awareness, which becomes especially relevant in the case of digital genres. This is so because presenters should pay attention not only to embodied actions (e.g., gestures, gaze and head moments) and the demands of the genre but also to the affordances offered by the digital medium.

5 Conclusion

This preliminary study has addressed the digital genre of 3MT presentations from both pedagogical and analytical perspectives. For this purpose, we first discussed a teaching approach to dealing with 3MT presentations. Then, we examined the presentations of two students, focusing on the organisation of rhetorical steps, their transmission of interpersonal meaning, and the range of semiotic modes that accompanied the expression of the latter.

Pedagogically, this study briefly described a multimodal genre-based approach to teaching 3MT presentations that have been implemented within an EAP workshop held at a Spanish university. The pedagogical proposal was devised considering that doctoral students should develop specific communicative skills to transmit complex content in a simplified and interactive way to non-expert audiences. The multimodal component was included to make doctoral students aware of the potentiality of using and combining varied semiotic resources to construct meanings, as well as to develop their literacy.

For the analytical part, we explored how two trained presenters organised discourse and expressed interpersonal meaning. The ELAN software supported the two analyses. As for the first analysis, we identified the moves and instances of interpersonal meaning that occurred in the 3MT presentations. Results showed that the two presenters followed the moves found by Hu and Liu (2018). Concerning the instances of interpersonal meaning, results revealed a greater number of them in the orientation, rationale and methods move. The use of the engagement marker "you" came to the fore frequently, shown to be used as a resource to create proximity with the audience. For the second analysis, a multimodal perspective was adopted to explore the variety of modes that the presenters instantiated while establishing interpersonal meaning. Findings indicated that the expression of engagement consisted of a variety of semiotic resources (e.g., facial expressions, body posture, gesture), which conformed to the multimodal ensembles. Overall, the two presenters made use of and combined semiotic resources to engage and create proximity with their audience. This multimodal representation points to a complex intersemiotic relationship (Jewitt et al., 2016) that, according to the findings of this study, was latent among speech, facial expressions, gaze direction, gesture and posture. The analysis suggests that the two presenters tended to make use of multimodal ensembles coherently and holistically to create meaning. This implies that the presenters were aware of the potentiality of using and combining diverse semiotic modes to achieve their communicative goals. In this case, sharing their research in an organised manner while simultaneously establishing interpersonal meaning with their audiences.

Further data would be needed to investigate the effect of the pedagogical model, as well as an analytical tool to examine performance from a multimodal lens. Regarding the second part of the study, a major concern was that the dataset was limited to two presentations. This was done because such a meticulous time-consuming analysis can only be done with a small representative dataset. Yet, further studies should expand the dataset to continue exploring how presenters express interpersonal meaning multimodally. In addition to this, for further research, it would be interesting to also analyse authentic presentations from a multimodal perspective to determine what characterises "good practices" in 3MT presentations. In turn, this would be beneficial for EAP teachers and trainers willing to prepare postgraduate students for 3MT presentations.

Acknowledgements This study has been funded by the Project UJI-B2020-09: Descripción y análisis multimodal de los repertorios genéricos docentes en contextos académicos de docencia en inglés en las disciplinas de económicas, ingeniería y ciencias de la salud.

References

Beltrán-Palanques, V. (2019). Multimodal pragmatics in FL interactions: The case of complaints and responses to complaints. In H. E. H. Lenk, J. Härmä, B. Sanromán, & E. Suomela-Härmä (Eds.), *Studies in comparative pragmatics* (pp. 179–198). Cambridge Scholar Publishing.

Beltrán-Palanques, V. (2022). Transitioning to live online lectures: An example based on elevator pitch presentations to foster ESP students' multimodal communicative competence. In J. R. Guijarro Ojeda & R. Ruíz Cecilia (Eds.), *Research and innovation in foreign language teaching: A global perspective* (pp. 297–317). Tirant lo Blanch.

Beltrán-Palanques, V. (2023). Multimodal literacy in ESP/EAP. Implications for professional development programs. In P. Hohaus & J. F. Heeren (Eds.), *Innovating the future of teacher education* (pp. 101–128) Brill.

Beltrán-Palanques, V., & Querol-Julián, M. (2018). English language learners' spoken interaction: What a multimodal perspective reveals about pragmatic competence. *System, 77*, 80–90. https://doi.org/10.1016/j.system.2018.01.008

Campoy-Cubillo, M. C., & Querol-Julián, M. (2015). Assessing multimodal listening. In B. Crawford & I. Fortanet-Gómez (Eds.), *Multimodal analysis in academic settings: From research to teaching* (pp. 193–212). Routledge.

Carter-Thomas, S., & Rowley-Jolivet, E. (2020). Three Minute Thesis presentations: Recontextualisation strategies in doctoral research. *Journal of English for Academic Purposes, 48*, 100897. https://doi.org/10.1016/j.jeap.2020.100897

Halliday, M. A. K. (1978). *Language as social semiotic: The social interpretation of language and meaning.* Edward Arnold.

Hardwood, N. (2005). "We do not seem to have a theory … The theory I present here attempts to fill this gap": Inclusive and exclusive pronouns in academic writing. *Applied Linguistics, 26*(3), 343–375. https://doi.org/10.1093/applin/ami012

Harinck, F., & van Leeuwen, E. (2020). *The art of presenting. Delivering successful presentations in the social sciences and humanities.* Cambridge University Press.

Hu, G., & Liu, Y. (2018). Three minute thesis presentations as an academic genre: A cross-disciplinary study of genre moves. *Journal of English for Academic Purposes, 35*, 16–30. https://doi.org/10.1016/j.jeap.2018.06.004

Hyland, K. (2002). What do they mean?: Questions in academic writing. *Text, 22*(4), 529–557. https://doi.org/10.1515/text.2002.021

Hyland, K. (2005). Stance and engagement: A model of interaction in academic discourse. *Discourse Studies, 7*(2), 173–191. https://doi.org/10.1177/1461445605050365

Hyland, K. (2006). Medical discourse: Hedges. In K. Brown (Ed.), *Encyclopedia of language and linguistics* (2nd ed., pp. 694–697). Elsevier.

Hyland, K. (2010). Constructing proximity: Relating to readers in popular and professional science. *Journal of English for Academic Purposes, 9*(2), 116–127. https://doi.org/10.1016/j.jeap.2010.02.003

Hyland, K., & Jiang, F. (2017). Is academic writing becoming more informal? *English for Specific Purposes, 45*, 40–51. https://doi.org/10.1016/j.esp.2016.09.001

Hyland, K., & Zou, H. (2021). "I believe the findings are fascinating": Stance in three-minute theses. *Journal of English for Academic Purposes, 50*, 100973. https://doi.org/10.1016/j.jeap.2018.06.004

Jewitt, C., Bezemer, J., & O'Halloran, K. (2016). *Introducing multimodality*. Routledge.

Lorés, R., & Diani, G. (2021). Disseminating knowledge: The effects of digitalised academic discourse on language, genre and identity. *European Journal of English Studies, 25*(3), 249–258. https://doi.org/10.1080/13825577.2021.1988262

Luzón, M. J. (2013). Public communication of science in blogs: Recontextualizing scientific discourse for a diversified audience. *Written Communication, 30*(4), 428–457. https://doi.org/10.1177/0741088313493610

Luzón, M. J., & Pérez-Llantada, C. (2019). Connecting traditional and new genres: Trends and emerging themes. In M. J. Luzón & C. Pérez-Llantada (Eds.), *Science communication on the Internet: Old genres meet new genres* (pp. 1–18). John Benjamins. https://doi.org/10.1075/pbns.308.01luz

Luzón, M. J., & Pérez-Llantada, C. (2022). *Digital genres in academic knowledge production and communication: Perspectives and practices*. Multilingual Matters.

Martin, J. R., & Rose, D. (2005). Designing literacy pedagogy: Scaffolding asymmetries. In R. Hasan, C. M. I. M. Matthiessen, & J. Webster (Eds.), *Continuing discourse on language* (pp. 251–280). Equinox.

McNeill, D. (1992). *Hand and mind: What gestures reveal about thought*. University of Chicago Press.

Morell, T., Beltrán-Palanques, V., & Norte, V. (2022). A multimodal analysis of pair work engagement episodes: Implications for EMI lecturer training. *Journal of English for Academic Purposes, 58*, 101124. https://doi.org/10.1016/j.jeap.2022.101124

Morell, T., Norte, N., & Beltrán-Palanques, V. (2020). How do trained English-medium instruction (EMI) lecturers combine multimodal ensembles to engage their students? In R. Roig-Vila (Ed.), *La Docencia en la Enseñanza Superior. Nuevas Aportaciones desde la Investigación e Innovación Educativas* (pp. 308–321). Octaedro.

Norte, N. (2018). The impact of multimodal ensembles on audio-visual comprehension: Implementing vodcasts in EFL contexts. *Multimodal Communication, 7*(2). https://doi.org/10.1515/mc-2018-0002

Querol-Julián, M., & Beltrán-Palanques, V. (2021). PechaKucha presentations to develop multimodal communicative competence in ESP and EMI live online lectures: A team-teaching proposal. *CALL-EJ, 22*(1), 73–90.

Querol-Julián, M., & Fortanet-Gómez, I. (2019). Collaborative teaching and learning of interactive multimodal spoken academic genres for doctoral students. *International Journal of English Studies, 19*(2), 61–82. https://doi.org/10.6018/ijes.348911

Qui, X., & Jiang, F. (2021). Stance and engagement in 3MT presentations: How students communicate disciplinary knowledge to a wide audience. *Journal of English for Academic Purposes, 51*, 100976. https://doi.org/10.1016/j.jeap.2021.100976

Rowley-Jolivet, E., & Carter-Thomas, S. (2019). Scholarly soundbites. Audiovisual innovations in digital science and their implications for genre evolution. In M. J. Luzón & C. Pérez-Llantada (Eds.), *Science communication on the Internet: Old genres meet new genres* (pp. 81–106). John Benjamins. https://doi.org/10.1075/pbns.308.05row

Wittenburg, P., Brugman, H., Russel, A., Klassmann, A., & Sloetjes, H. (2006). ELAN: A professional framework for multimodality research. In *Proceedings of LREC 2006*, Fifth International Conference on Language Resources and Evaluation (pp. 1556–1559).

11

Introducing Science to the Public in 3-Minute Talks: Verbal and Non-verbal Engagement Strategies

Juan C. Palmer-Silveira and Miguel F. Ruiz-Garrido

1 Introduction: Science for Everyone

For many years, science has shown the image of being just for researchers and scientists (Dubos, 2013; Greene, 2013). Science, as such, was almost a forbidden word for the rest of the population, as there was an aura of mystique involving what scientists and researchers were doing (Calsamiglia, 2003; Calsamiglia & van Dijk, 2004; Massarani & Moreira, 2004). It was hard to find scientists interested in showing what they were doing to the general public. Alcíbar (2004) exemplifies the story of Joseph-Jérôme Lefrançois de Lalande, who directed the Astronomical Observatory in Paris at the end of the eighteenth century. Lalande used to bring one telescope to one of the main bridges over the Seine and showed people all the stars in the sky, explaining to them some basic concepts on astronomy. In fact, his house became an astronomical seminary. His main concern, therefore, was to approach science to everyone.

J. C. Palmer-Silveira • M. F. Ruiz-Garrido (✉)
Universitat Jaume I, Castelló, Spain
e-mail: palmerj@uji.es; mruiz@uji.es

R. Plo-Alastrué, I. Corona (eds.), *Digital Scientific Communication*,
https://doi.org/10.1007/978-3-031-38207-9_11

Fortunately, and since the turn of the twenty-first century, there has been a trend trying to show that science can be grasped by the layman. In fact, the attempt of popularisation of science is nothing else than "an endeavour to image scientific ideas in such a way that everyone (especially non-scientists) can grasp the fundamental concepts and have an idea of what science in essence is" (Cornelis, 1998, p. 30). That interest in learning basic knowledge on science can be seen nowadays, as data show that many people are interested in learning about science by the use of different formats (magazines on science, TV programmes, web pages or YouTube channels) (Camus, 2009; Takahashi & Tandoc, 2016). In any case, the way science is often offered throughout mass media has not been studied in detail, and "there has been very little study of the relationship between TV schedulers, directors, producers, managers, and screenwriters with science and scientific content" (Soto-Sanfiel et al., 2022, p. 228). It is necessary, therefore, to learn how to improve the public speaking abilities among scientists, in order to offer their research in a simple and appropriate way that could be understood by the maximum number of people (Youknovsky & Bowers, 2020; Xia & Hafner, 2021).

Academics have been aware of this new trend of making science available to the public; some authors use the expression "the third wave of studies" to talk about the dissemination of science via the Internet (Rowley-Jolivet & Carter-Thomas, 2019, p. 82). The result is a growing acceptance of online genres (i.e., parascientific genres) as tools to democratise science and spread scientific research to everyone (Luzón & Pérez-Llantada, 2019). This has caused two main outcomes: first, the appearance of academic genres that change their medium of dissemination without any adaptations, slight adaptations or huge changes; second, the continuous emergence of new genres, hybrids and/or specifically web-native genres. This evolution also implies that the interest for the dissemination and popularisation of science sometimes imply infotainment/edutainment (e.g., Marinov, 2020; Niemann et al., 2020).

Observing these changes, some of those new genres have derived into brief presentations of scientific issues for a heterogeneous audience (Carter-Thomas & Rowley-Jolivet, 2020). Recently, this type of short talks has even generated some competitions, especially devised to communicate science to the masses, focusing on offering an attractive delivery either to specific (Agyapong et al., 2019) or general audiences. Many academic institutions have devoted efforts to organise this type of science popularisation competitions. Among those events, the 3MT (3-minute thesis competition), originally developed by the University of Queensland, is probably the best known one within the academia, and some studies have analysed its features (Hu & Liu, 2018; Qiu & Jiang, 2021; Jiang & Qiu, 2022). Nevertheless, there are other options of science dissemination that deserve some attention, being them either competitive (such as FameLab or Science Slam) (Scalice et al., 2019; Stimm, 2020) or simply just for fun (like Big Van Science) (González Burón & Marimón Garrido, 2021).

2 FameLab, a Competitive Event

In this study, we will focus on the FameLab scientific talks international competition. The reason to choose this format is based on the support given by the British Council to their founders, the Cheltenham Science Festival. This event has been organised since 2005, first on a national basis and, some years later, on an international scale. It is held in different countries of the world, and the winners of each country go through a series of qualifying events until a final session, with the ten best science communicators gathering in a final event, in which different prizes are given. To the best of our knowledge, this type of presentation has not been studied yet, and certainly not from a multimodal standpoint.

The bases of the talks are clear: contestants should offer a 3-minute talk on scientific and/or technological subjects, followed by a brief conversation with members of the jury. Props are allowed if competitors can take them on stage and set them up very quickly; additionally, PowerPoint presentations (or any other similar system of visual support) are not allowed. This makes a huge difference with traditional presentations of research communication such as TED talks or seminar and conference presentations. Hu and Liu (2018, p. 21) point this out, stating that "rules regarding the use of slides and props separate it [3MT] from other similar academic research communication genres (e.g., conference presentations, TED talks) and other science communication competitions (e.g.,

FameLab)". Additionally, there are other remarkable differences, such as the fields of scope of the research presented (FameLab is only focused on the STEM field).

The contestants should be young (in their late 20s or early 30s), working as researchers in academia, or in public or private institutions. The main objective of this type of event is "to identify, mentor, and link young talented science communicators" (Zarkadakis, 2010, p. 281). In fact, the organisers show a clear commitment to finding ways in which scientists could enter the public domain. It follows the format of talent TV competitions, such as Got Talent or similar ones. As it is the case in this type of shows, the FameLab setting traditionally comprises a stage, the competitor(s), a jury (from academia and the media), and a live heterogeneous audience. Additionally, the vital element of this event is to enhance knowledge throughout entertainment, as it should be "a pleasant and exciting show that presents concepts and ideas from science in a delightful way, by new 'faces' of science that upturn the usual stereotypes" (Zarkadakis, 2010, p. 282).

Being a competition, it is important to know the criteria followed to get the winners of this event. According to the website of the contest, and from its inception (Cheltenham Science Festival, 2022), the panel follows three criteria to choose the winner, known as the 3 Cs: content (simple structure and creative narrative), clarity (bringing the topic to life in a simple, understandable way) and a third element, charisma, that deserves further study, as some people understand the term with different perspectives (Tubbs, 2019).

Riggio (1998) states that there is no generally agreed-upon definition of charisma or, as Nikoloski (2015, p. 18) points out, "any attempt to pin it down to a specific set of human qualities would be futile". For almost one century, authors have noticed that, apart from content, science is based on the charisma of the person explaining that piece of research to her/his counterparts (Parsons, 1937; Shils, 1965, 1968; Gustin, 1973), and this feature is even more relevant when offering information to lay people. In the webpage of the festival, charisma is defined as that "hard-to-describe but unmistakable quality" that makes audience and judges be left inspired and enthused about science. It is a question of putting together passion and communication, making science "easy to listen to, entertaining and exciting" (Cheltenham Science Festival, 2022).

Some features that a charismatic speaker should possess are defined by Reynolds (2013) when he states that this type of communicators, apart from being thoroughly prepared and get direct connections with their audience, should use an open body posture (e.g., use their hands while speaking), and be passionate, energetic and optimistic (moving around can help). Additionally, they should also speak in a clear, fluent, forceful and articulate way, as well as constantly alter the intonation and pacing of their delivery. The use of all these elements, already observed by Wiseman (2008), are a necessary combination to engage the audience. As Kirova (2020) comments, presentations should, therefore, communicate on two levels: verbally and non-verbally, but it is the combination of both levels that will force listeners to be persuaded and engaged by the speaker.

3 Engagement

All the efforts that communicators carry out in FameLab presentations are based on engaging the audience, although engagement is not an easy concept. As Mersey, Malthouse and Calder point out (2010, p. 40), "practitioners and scholars do not agree on a definition of engagement. If we cannot define engagement, we cannot measure it." Nevertheless, everyday connotations of engagement, as Schaufeli (2013, p. 29) states, refer to "involvement, commitment, passion, enthusiasm, absorption, focused effort, zeal, dedication, and energy". For our purposes, engagement will be the ability that speakers will have to attract the audience through involvement and passion, forcing the active participation of the audience, responding to their concerns, and leading them towards intended interpretations and conclusions. Or, as Hyland states, "recognizing their uncertainties, including them as discourse participants, and guiding them to interpretations" (Hyland, 2010, p. 111).

Presenters often think of ways to make their presentations as engaging as possible, considering the time constraint (3 minutes) they must face, and analyse how to help the audience to focus on them rather than on any other things happening in the room (Lee et al., 2013). Unfortunately, as Stuart (2013, p. A4) states, "the training in audience engagement is not part of the typical education of a scientist. The rare speakers who can truly hold an audience in the palm of their hand have usually learned their skills through having been mentored at some point and then by rehearsing their talks, often with colleagues." The presentations selected in our dataset have reached the final stages of the contest due to their high level of engagement, having received their own national award. As Martin and Bolliger (2018) point out, the concept of engagement has a lot to do with interaction, and both terms are often used interchangeably.

There are many definitions of this concept, but it is probably Johnston (2018) who underscores the duality between the psychological process implied in the term and the need to reach an aim which, in many instances, is quite tangible. In Johnston's opinion (2018, p. 19), engagement is "a dynamic multidimensional relational concept featuring psychological and behavioural attributes of connection, interaction, participation, and involvement, designed to achieve or elicit an outcome at individual, organisation, or social levels". From a discourse analysis standpoint, engagement is based on "the interpersonal and rhetorical choices that writer/speakers make to attract and hold interest and command attention". (Hyland & Zou, 2021, p. 25). Studying academic 3-minute presentations, those authors propose five devices that tend to be included in any piece of discourse in order to get the listeners' attention: hearer mentions, directives, questions, appeals to shared knowledge and personal asides. Additionally, Carter-Thomas and Rowley-Jolivet (2020), also analysing 3-minute talks, include other strategies to engage the audience, such as the use of titles, visual impact, personalisation devices, or what they call humour and "street cred". As it can be seen, the multimodal component of the presentations is also considered in those latter strategies (visual impact). Thus, the combination of some verbal and non-verbal features—that is, the use of multimodal resources or strategies—are also part of spoken discourse analysis, and they become a useful tool of communication to engage the audience (e.g., Bateman, 2019; Valeiras-Jurado & Bernad-Mechó, 2022).

4 Aim of the Study

Our observation of FameLab talks reaching the final has suggested that they could be taken as examples of good practices, showing a positive, assertive attitude while delivering a message, which seems to be an important element in those presentations. In this study, our aim is to identify the verbal and non-verbal features participants use to present content and engage the audience. We aim at looking at the existence of some multimodal features that may be intended to involve the audience, compelling them to react to the presentation.

5 Methodology

In this study, we analyse the 2020 FameLab talks that reached the final event of the competition, offered by ten international young scientists. All these presentations have gone through a series of qualifying stages before reaching these finals. This implies that these talks will probably show a significant level of engagement, as they have been tested on several prior occasions by the presenters, adapting their delivery to a very special setting. This time and due to the pandemics, they were not delivered live in front of a physical audience, changing their delivery process from their regular one.

All ten talks studied were broadcast online from the participants' home countries throughout a live event. Due to this special situation, with just one steady camera in front of the speakers, the analysis will not depend on the camera shots (unlike what happens in similar genres that use filmic modes; see Valeiras-Jurado & Bernad-Mechó, 2022) or on the audience's reactions, as they could not feel those reactions in front of them (opposite to what happens in other online science dissemination genres). As it is a competition, these young scientists were aware that live viewers could vote for their favourite presentation from home.

The analysis of the engagement features within our dataset has followed two main works, Hyland and Zou (2021) and Carter-Thomas and Rowley-Jolivet (2020) studies. However, our observation of the presentations gathered have led us to eight engagement strategies that we consider relevant enough to be analysed. Thus, our proposal is the following one:

1. Hearer mentions
2. Directives
3. Questions
4. Appeals to shared knowledge
5. Visual impact
6. Aural impact
7. Personalisation strategies
8. Humour and street cred

As for the structure of the FameLab presentations, which is not our aim in this piece of research, we will be referring to initial or final sections to explain the use of some features.

The total amount of words used in the ten presentations analysed is 4536. Considering that the time regulations imply that these talks can only be 3 minutes long, the length is quite similar, as we can see in Table 11.1.

6 Results and Discussion

The listed features commented above used by speakers to engage the audience have been analysed, and the results show that all play a role in the ten presentations forming our dataset.

6.1 Hearer Mentions

Hearer mentions means explicit ways to refer directly to the public. We have observed that connection with the audience throughout the use of "you/your" that appears 78 times in the ten presentations, and at least once in all cases (just one case in P9 and two in P10). From this amount of second-person references, 61 of them are personal pronouns, and we

Table 11.1 Dataset description

Presentation	Country	Speaker	Field of expertise	Length of the talk	Number of words
P1	Australia	Cody Frear (Sp1)	Medicine	3.20	510
P2	South Korea	Gayun Bu (Sp2)	Molecular biology	4.26	513
P3	Great Britain	Rebecca Ellis (Sp3)	Medicine	2.50	462
P4	Switzerland	Sauradeep Majundar (Sp4)	Chemistry	2.53	460
P5	Malaysia	Atteleth Don Peris (Sp5)	Marine science	2.54	474
P6	Kazakhstan	Shokhista Ismailzhanova (Sp6)	Medicine	2.48	346
P7	Brazil	Gabriela Ramos Leal (Sp7)	Veterinary medicine	2.50	382
P8	Qatar	Ahmed Maani (Sp8)	Cryptography	3.07	428
P9	South Africa	Pinky Mokwena (Sp9)	Environmental sciences	2.59	414
P10	Egypt	Mahmoud Beshir (Sp10)	Medicine	3.28	547

Table 11.2 Number of appearances of the pronoun "you" and functions implied in its use

Functions	Number of occurrences	Percentage (%)
Shared experience	4	6.56
Guiding understanding	19	31.15
Inviting interpretations	17	27.87
Thanking audiences	9	14.75
Creating dialogue	12	19.67
	61	100.00

have followed Hyland and Zou's taxonomy (2021) to point out the functions they imply in our dataset, observing a varied usage, as depicted in Table 11.2. To complete those functions we have added Fernández Polo's (2018) category *Creating dialogue: inviting questions and comments.* Although questions are explained below, the use of the pronoun "you" within questions required a particular analysis, as we have seen that it becomes relevant in our dataset, since six presenters use it, probably trying to get as close to their audience as possible, even knowing that they are unfortunately not there during the presentation itself. In this dataset we have also observed that the percentage of use of this pronoun as a way to share experience with the audience is considerably lower to the results observed in Hyland and Zou's study (2021). In our dataset, presenters probably understand that this function does not contribute much to engage an unknown and unseen audience, and they just try to guide their understanding and invite them to make their own interpretations.

Most young scientists intend to get close to the audience using this second-person pronoun, trying to create some direct connection between speakers and listeners. Additionally, in all cases but one (Sp6) these scientists use "thank you" as a final expression to conclude their 3-minute presentations. In the case of the exception, the presenter gets close to the camera and, instead of greeting the audience, seems to shake hands with them, as she has decided to perform her talk playing the role of the Coronavirus (the main topic of her talk); as a result, she is trying to shake our hands "virtually", with the purpose of showing how simple may be to infect us.

Another way of integrating the audience within the presentation is by using other personal pronouns, such as "we", "our" or "us". In our dataset, there are 34 occurrences of "we" in eight presentations, 23 of "our" in six presentations, and 12 of "us" in six presentations. Among the examples of "we", over half of them (18) are highlighting the role of the researchers, where 16 occurrences promote the audience's involvement and make them participants/researchers. Concerning "our", the results are quite different, as only 6 occurrences exclude the audience, but 17 make them feel included. Finally, "us" confirms the previous tendency, since ten occurrences try to integrate the audience, leaving two examples regarding only researchers. These examples may prove that most speakers are quite aware of the need to make the audience be part of the speech and of the event and actions explained.

Using hearer mentions often implies an attempt to enhance persuasiveness: showing the audience how much the speaker cares about them, paying attention to the hearers' interest on the topic and involving them in the piece of research summarised in the three-minute talk. It tends to be used to share the experience with the hearers, but also to guide them in the scientific process developed and in the observation of the results achieved.

6.2 Directives

The second feature observed in our dataset is directives, which are commands used by the speaker to recommend the members of the audience to act in a specific way. They are, therefore, often connected with that direct appeal observed when using second-person pronouns. In this case, the presenter uses them to let us know how to act during her/his talk. The total number of directives observed in our dataset is 31, from which most of them are imperative sentences (22), while the amount of obligation modals is considerably lower (only 9 cases in our dataset). In any case, not all presentations use directives (just seven out of ten). In the examples collected, those directives tend to be imperative verbs, though the use of obligation modals has also been detected. Once again, this is a common feature in most presentations analysed, and in some cases those commands appear in pairs, instructing them how to perform some actions, as in examples 1 and 2.

(1) ***Bust out your sweat bands*** *and* ***loosen up your limbs*** *because you are about to be immersed in one of the greatest competitions of all time.* (P1)
(2) *Hello, everyone. As you're watching me across your screens,* ***take a moment*** *and* ***have a look*** *at all the materials around you.* (P4)

6.3 Questions

Questioning the audience is another simple way to get in touch with the listeners and make the talk more appealing, engaging them throughout a direct connection with the speaker. This often enhances dialogic engagement in traditional face-to-face presentations, but our dataset is formed by online talks. In our dataset, there are 27 rhetorical questions appearing in eight presentations, used to make people think about the validity of the study presented. In 14 cases, those questions appear in the introduction of the talk, in those initial questions in which the speakers establish the topic of their presentations (samples 3, 4 and 5).

(3) ***Can you recall the last time you were captivated by nature?*** (P5)
(4) ***What if I told you 2020 has been a fantastic year?*** (P6)
(5) ***Have you ever researched to stop the time?*** *Maybe to keep a sweet memory or to freeze a precious moment.* ***It sounds like fantasy, right? But what if I told you that it can be done in real life?*** (P7)

Similarly, the rest of the questions (13) are also rhetorical ones, used to establish ideas in a more appealing way. This is mainly carried out along with the presentation (11 cases) (example 6), but we should point out that it has also become a resource used by some scientists to offer their final concluding remarks in two presentations (example 7).

(6) ***How can we improve outcomes for children like Jack?*** (P1)
(7) ***So tell me, do you like the cold?*** (P7)

6.4 Appeals to Shared Knowledge

In scientific articles, prior knowledge is basic to generate the basis of the piece of research (Sun & Linton, 2014; LaPlaca et al., 2018). This is something that presenters may know, and they are aware that the audience that watches their presentations are not necessarily experts in the field. That is why they may have decided, in the initial part of their presentations, to establish some background knowledge that will be used as a starting point for what is about to come. We have observed examples of two different possibilities to engage the audience by offering references to previous research and knowledge.

Initially, we have seen that a presenter can introduce her/his talk by simply establishing a (well-known) case, something that anyone can understand, as it is widely known by the laymen. It is a starting point based on general knowledge and common sense.

(8) ***Hi, everyone. The fight against the cancer is challenging, but one of the greatest obstacles is cancer's ability to resist treatment. And to combat this, scientists have developed new techniques and new drugs that target specifically the cancer cells.*** (P10)

A second option would be to introduce information on historical events that are basic to understand what is about to be presented. This option is used to establish a point of departure, a milestone, on which to build the piece of discourse to be delivered. Using history to start a presentation is often used in business presentations (Burton & Obel, 1995; Fitzpatrick, 2017), but it is probably not so common in scientific discourse, though the need to support research on previous studies is a must for any young scientist.

(9) ***It's the year 1943. The UK is still struggling in the war against the Nazi, but somewhere in an English country house, Alan Turing and his friends are developing the bar machine, the machine which (…) will help them win the war.*** (P8)

6.5 Visual Impact

Regarding visual impact, these young scientists tend to use several features to get closer to the audience and engage them into their brief presentations. As explained earlier, there is only one camera shot, so there is no possible editing of the video or any filmic modes to apply (Valeiras-Jurado & Bernad-Mechó, 2022). Considering that there are some limitations in this issue (no PowerPoint slides can be used but easy-to-use props are possible), this category refers to any element (external to the speaker or part of their body language) that can support speakers in making their talks more attractive and engaging. Thus, we include diverse aspects such as eye contact, hand movements with diverse functions, both part of speakers, and prop use, background and garments, as external resources.

Speakers' Body Language

1. Eye Contact

As presentations are being broadcast via the Internet, there is an initial feature that all the speakers use, which is direct eye contact with the camera. In all cases, the presenters try to keep eye contact, knowing that it is a basic feature to keep the interest of the audience. They all look directly at the camera, trying to replicate what they would do in front of a live physical audience. Additionally, they also opt for smiling directly at the camera, trying to create a positive rapport with the audience. However, those smiling faces require some additional comments. In two of the cases, Sp1 and Sp2 stop smiling during the concluding remarks, to stress the importance of their results (Sp2 is shown in Fig. 11.1, above). Furthermore, Sp5 also uses an exaggerated gesture: while talking about a rotten egg, he frowns to show displeasure (see Fig. 11.1, below). It is obvious that these differences in the gestures chosen make a difference in the way the audience understands the message.

Fig. 11.1 Examples of smiles/frowns

2. Hand Movements

A second visual element analysed is the use of hand movements in order to stress concepts and force the audience to pay attention to what speakers are about to point out. Hand movements vary among the ten scientists, but it is something all of them use at different levels. Figure 11.2 shows some examples of hand movement to engage the audience.

Nevertheless, hand movements go even further when some subjects analysed even use one pointing finger to stress some concepts or to force the audience to pay attention to a specific feature of their piece of discourse (see Fig. 11.3). In some cases, fingers are also used to establish the order of appearance of concepts in a specific moment of the talk (*first, second, third...*), as in the third image on the right.

Additionally, we have also observed that, in a couple of examples, and in order to conclude the talk, the presenters use their hands to wave goodbye to the audience, either by (virtually) shaking their hands (left image in Fig. 11.4) or by completing a military salutation (right image in Fig. 11.4). In both cases (Fig. 11.4) they are trying to do something different and remarkable to finish the presentation in a positive way.

Fig. 11.2 Examples of hand movements

Fig. 11.3 Examples of pointing fingers

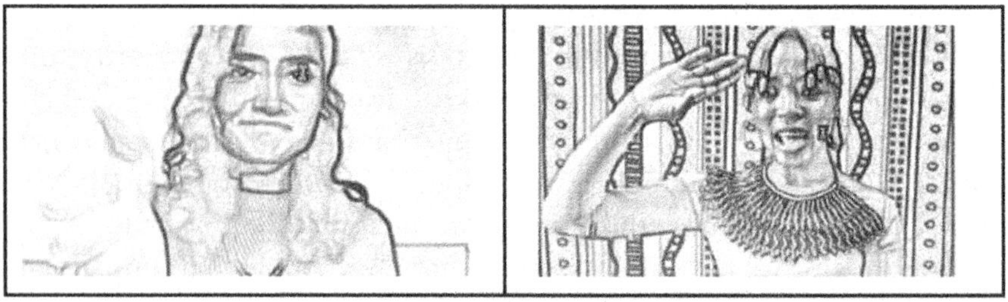

Fig. 11.4 Examples of gestures to conclude the presentation

Speakers' External Resources

1. Props

Apart from the way the scientists analysed use gestures to convey the message, most of them also take advantage of the opportunity that organisers give them to bring props to increase the level of engagement with their audience (as we have explained earlier, props are the only element allowed in the contest). They are a simple way to do something different, unexpected, and they are often used to hook the audience, but also to involve them in specific moments of the presentation. In the dataset, we have seen that six young scientists have taken advantage of this possibility, and they bring different objects. For instance, Sp1 brings a rugby ball to introduce a moving story, whereas Sp2 used paper grippers to explain a physiological process (both examples in Fig. 11.5 above), and Sp9 shows a Cupid's arrow to establish a comparison between love and the way carbon and chlorine combine, showing later some powder as a result of her study (bottom-left image in Fig. 11.5). Additionally, Sp6 wears a crown, playing the role of *Miss Coronavirus*. Much more traditional is the type of

Fig. 11.5 Examples on the use of props

material shown by Sp4, a model of the cells and some atoms and molecules that the presenter takes out from his pocket. Finally, Sp8 shows a constant (and maybe excessive) use of props, especially devised for the occasion: a cardboard replica of a TV set, cardboard signposts, or cards with concepts, among others (bottom-right image in Fig. 11.5).

We would like to point out a specific case: Sp2 plays a role during the first two thirds of the presentation (becoming a protein) and then, in the final part, she just becomes a human being by closing the camera eye for a second and taking off her glasses. From that moment onwards, she takes the role of a young Korean researcher, and her image without the glasses makes it visible to the audience that we are now dealing not with a protein, but with a person (Fig. 11.6). This allows the speaker to perform two different roles, making the presentation more appealing to the audience.

2. Set Background

Another important decision, considering the online performance, is to choose the right background and the perfect outfit. Regarding the first

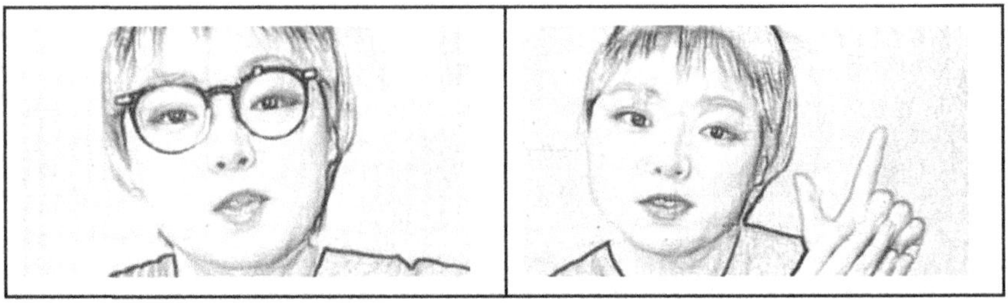

Fig. 11.6 Specific use of a prop to define two different characters within a presentation

Fig. 11.7 Use of background settings in a presentation

concept, one half of the researchers have opted for non-textured walls, mostly white, but there are a few exceptions that have forced us to analyse the importance of choosing the right background setting. For instance, textured walls appear in P5, P8 (Fig. 11.7, top-right and bottom-right images, respectively) and P9, with different colours and designs, always combined with the way speakers are dressed. On the other hand, P1 is based on the appearance of an image in the background, an Olympic stadium, as Sp1 talk includes some elements based on sport (Fig. 11.7, top-left image). The most different background is that used by Sp4, who delivers his presentation from his laboratory, and we can see a more traditional, science-related background (Fig. 11.7, bottom-left image). It seems that, in this case, the choice of using the natural environment for his daily activities has helped him to feel at ease and comfortable.

3. Garments

Choosing the garments that can offer a better image in front of an audience is a relevant aspect for many researchers, as the viewers of this event tend to pay attention to these details, and a professional outfit may increase our professional perception. In any case, the use of outfits to introduce the presentations can be defined as varied: they range from T-shirts (Sp7; Fig. 11.8, top-right image) or informal/sports outfit (Sp4; Fig. 11.8, bottom-right image) and tracksuit jackets and jeans (Sp10; Fig. 11.8, top-left image) to ethnic dressing (Sp9; Fig. 11.8, bottom-left image). Ties (Sp1) and even bow ties (Sp10) are also used. Everyone has chosen the look that they thought better suits their presenting task and, probably, even their personality (Fig. 11.8).

Fig. 11.8 Use of different outfits

Additionally, most presenters have chosen to stand up in order to offer their talks. Some stand in the same position all the time, such as Sp1 and Sp5, while others move back and forth to point out ideas and facts, such as in Sp3 and Sp7, in a more dynamic approach to the presentation.

6.6 Aural Impact

Analysing our dataset, we found a feature not included in previous literature on the topic because it is mainly a multimodal feature. During our study, we noticed that the use of the right tone of voice in order to deliver a message is a key factor to convey a message to the audience, and this is even more relevant when dealing with an online presentation, as the sound must be loud and clear. Prosody becomes a relevant element to increase the level of engagement that most presenters try to acquire in their talks. Pace and rhythm become an important element in the delivery of these sessions.

Using different tones during the talks, by combining different levels of pace and rhythm, allows these young researchers to offer a brief summary of the activity to the audience, highlighting certain ideas or concepts. Considering that English is not the native language in most cases, it is important to find the perfect pace that allows them to carry out this task successfully. In any case, all the presenters have tried to play with prosody to engage their audience. They have changed their tone and rhythm, in an attempt to stress important ideas, trying to convey a positive approach by playing with their voice, such as Sp4, Sp5 or Sp7.

Prosody is an integral part of oral communication, but in some situations the presenters go even further. In some cases, the speakers even yell at the audience as a way to call their attention. For instance, Sp7 plays a lot with her voice, even shouting one word (*stop*). Similarly, Sp10 changes the tone he is using during his presentation, yelling twice and moving his body and hands in an amplified way. It is important to analyse that those two moments when the presenter yells are accompanied by physical movements that break the usual pace and rhythm chosen by him, probably trying to get our attention as viewers. This is an example of how potent combining verbal and non-verbal engagement strategies can be.

Additionally, Sp9 even fakes a beeping sound to imply an alert. Finally, Sp1 uses some background music (the Olympic Hymn) to create an impact on the audience at the beginning of his presentation.

6.7 Personalisation Strategies

There are two cases where personalisation plays a dominant role: in both examples, the presenter takes a role in order to introduce the presentation (Sp2 and Sp6). In the former, the speaker plays the role of a protein, and in the latter, she becomes the coronavirus. Both examples combine not only the personalisation of the speaker as such, but also a different outfit (glasses or a crown, respectively).

Nevertheless, in some other cases these young researchers have not gone so far, but they have included personalisation as a device to introduce and clarify some aspects of their talks. Using stories and anecdotes tends to help the speaker to exemplify concepts, and that, in turn, facilitates the comprehension of the message. In our dataset, this feature has been used by five researchers. In some cases, they opt for emotional stories (Sp1 introduces the story of Jack, a rugby player, and Sp9 tells the story of a friend's grandparents, who celebrated 61 years of marriage) or popular knowledge (Sp7 depicts how cold works by explaining the story of Captain America). Nevertheless, in other cases this use of storytelling becomes paramount, as Sp8 introduces historical data on the works of Alan Turing and Al-Kindi, or as in the case of Sp6, in which the whole presentation is a story in which she has become the main character, using the first person in order to introduce it, explaining how she (the COVID-19 virus) invades the lungs.

Finally, some presenters have used comparisons to make things simpler for the audience, trying to clarify concepts that might be difficult for those not working in the specific field of study. In our dataset, Sp1 compares athletes and cells, Sp2 does the same with humans and molecules, and Sp10 establishes connections between a community and our body and between architects and fibroblasts.

6.8 Humour and Street Cred

Traditionally, audiences tend to enjoy presentations offered from a positive perspective. Speakers often use engagement strategies as jokes, parenthetical remarks or asides to imply closeness and intimacy with the audience (Oliveira et al., 2021). That is why the use of humour tends to be a must in many instances. In our dataset, it seems there is a positive approach to communication in all the participants, but in some instances, they even opted for hints of humour in their speeches, ranging from puns and polysemy to darker humour and stereotypes.

(10) P1 (pun): ***the U-skin Bolts, if you wish***
(11) P2: (playing with misconceptions and stereotypes) ***I am a woman, Korean and nerd… But maybe I am stronger, bad at maths and not so nerdy.***
(12) P6: (dark humour) ***2020 was a great year for me, Coronavirus.***
(13) P8: (polysemy) ***Cribs… well, other type of cribs.***

We should keep in mind that the use of humour helps to offer a positive approach to presentations and this, in turn, enhances the overall ability of speakers to engage the audience in the talk (Riesch, 2015; Ruiz-Madrid & Fortanet-Gómez, 2015).

7 Conclusions

In this FameLab talks, discourse cannot be analysed in an isolated way: the fact of having to summarise all the information in just three minutes forces these young researchers to offer information in an active and thought-provoking way, as the instructions of the contest requires (reach a varied kind of audience in an entertaining way through the proper and accurate content, the clarity of delivery and the charisma of the speaker). Our analysis has shown the high amount of engagement features used by speakers: some of them basically verbal but some others non-verbal, both playing an attractive role for the audience by themselves but increasing their potential effect when appearing in combination. Thus, multimodality plays an important role, not only in engaging the audience in a complex scientific topic but also in transmitting knowledge as part of the same event. As the instructions offered by the organisers make clear that visual aids are limited (just props), verbal and non-verbal communication combined become more relevant.

Through the study, we have tried to identify the features used in this kind of infotainment/edutainment presentations, as nothing has been done to our knowledge. However, our study also shows some limitations. We have not paid attention to the exact quantity of some engagement strategies used in each presentation or in combination, as it was not our aim to develop a quantitative analysis. Some of our results have been explained from a multimodal standpoint as a whole. Needless to say, a more accurate multimodal study, looking at the effectiveness of both verbal and non-verbal communication together, is required in the future.

We should consider that in an ever-changing world, where the Internet has helped us to overcome many communication problems (as the ones we have had since March 2020), disseminating scientific contents can be enhanced thanks to this type of new emerging genre. Science has always evolved hand in hand with technology, and this is just one good example on how the Internet can be used to implement the possibilities of communicating science in an international setting. Technologies are here not only to stay, but to constantly evolve. As ESP practitioners, we hope that our research could bring new rhetorical tools to be applied in our teaching, but also the understanding of how science is communicated broadly and how scientists' discursive practices evolve and get adapted to a new, dynamic reality. Additionally, we also consider that, as it happened with other types of discourse, hard sciences have just opened the way, and that this type of competition will soon be used in soft sciences too in order to disseminate knowledge.

Acknowledgements This study has been supported by two projects: Project PID2021-127827NB-I00 funded by MCIN/AEI/10.13039/501100011033 and ERDF A way of making Europe, and project CIAICO/2021/069 funded by Generalitat Valenciana–Conselleria d'Innovació, Universitats, Ciència i Societat Digital. We thank FameLab organisers (Cheltenham Festivals) for their collaboration and allowing us to use their data.

References

Agyapong, V. I. O., Hrabok, M., Agyapong-Opoku, G., Khinda, H., Owusu-Antwi, R., Osei, A., Ohene, S., Ulzen, T., & Gilligan, P. (2019). Evaluating the impact of an innovative public speaking competition to promote psychiatry as a career option for Ghanaian medical students. *Academic Psychiatry, 43*, 180–183. https://doi.org/10.1007/s40596-018-0986-3

Alcíbar, M. (2004). La divulgación mediática de la ciencia y la tecnología como recontextualización discursiva. *Anàlisi, 31*, 43–70. https://raco.cat/index.php/Analisi/article/view/15153

Bateman, J. A. (2019). Information design and multimodality: New possibilities for engagement across theory and practice. *Information Design Journal, 25*(3), 249–257.

Burton, R. M., & Obel, B. (1995). *Strategic organizational diagnosis and design*. Springer.

Calsamiglia, H. (2003). Popularization discourse. *Discourse Studies, 5*(2), 13–146.

Calsamiglia, H., & van Dijk, T. A. (2004). Popularization discourse and knowledge about the genome. *Discourse Society, 15*(4), 369–389.

Camus, J. T. W. (2009). Metaphors of cancer in scientific popularization articles in the British press. *Discourse Studies, 11*(4), 465–495.

Carter-Thomas, S., & Rowley-Jolivet, E. (2020). Three-minute thesis presentations: Recontextualisation strategies in doctoral research. *Journal of English for Academic Purposes, 48*, 1–14. https://doi.org/10.1016/j.jeap.2020.100897

Cheltenham Science Festival. (2022). FameLab rules. Retrieved March 21, 2022, from https://www.cheltenhamfestivals.com/science-/famelab/rules/

Cornelis, G. C. (1998). Is popularization of science possible? *Paideia, 37*, 30–33. https://doi.org/10.5840/wcp20-paideia199837647

Dubos, R. J. (2013). Science and the layman. In I. B. Cohen & F. G. Watson (Eds.), *General education in science* (pp. 3–15). Harvard University Press. (Original work published in 1952). https://doi.org/10.4159/harvard.9780674283596

Fernández Polo, F. J. (2018). Functions of "you" in conference presentations. *English for Specific Purposes, 49*, 14–25.

Fitzpatrick, E. (2017). *Persuade on purpose: Create presentations that influence and engage*. Mercier.

González Burón, H., & Marimón Garrido, O. (2021). Von Big Bang zu Big Van. In M. D. Weitze, W. C. Goede, & W. M. Heckl (Eds.), *Kann Wissenschaft witzig?* Springer. https://doi.org/10.1007/978-3-662-61582-9_9

Greene, A. E. (2013). *Writing science in plain English*. University of Chicago.

Gustin, B. H. (1973). Charisma, recognition, and the motivation of scientists. *American Journal of Sociology, 78*(5), 1119–1134.

Hu, G., & Liu, Y. (2018). Three-minute thesis presentations as an academic genre: A cross-disciplinary study of genre moves. *Journal of English for Academic Purposes, 35*, 16–30. https://doi.org/10.1016/j.jeap.2018.06.004

Hyland, K. (2010). Corpus informed discourse analysis: The case of academic engagement. In M. Charles, D. Pecorari, & S. Hunston (Eds.), *Academic writing: At the interface of corpus and discourse* (pp. 110–128). Continuum.

Hyland, K., & Zou, H. J. (2021). Pithy persuasion: Engagement in 3-minute thesis presentations. *Applied Linguistics, 43*(1), 21–44. https://doi.org/10.1093/applin/amab017

Jiang, F. K., & Qiu, X. (2022). Communicating disciplinary knowledge to a wide audience in 3MT presentations: How students engage with popularization of science. *Discourse Studies, 24*(1), 115–134. https://journals.sagepub.com/doi/10.1177/14614456211037438

Johnston, K. A. (2018). Toward a theory of social engagement. In K. A. Johnston & M. Taylor (Eds.), *The handbook of communication engagement* (pp. 19–32). Wiley Blackwell.

Kirova, A. V. (2020). Verbal and nonverbal means of persuasiveness in business presentations. *Neophilology, 6*, 41–48. https://doi.org/10.20310/2587-6953-2020-6-21-41-48

LaPlaca, P., Lindgreen, A., & Vanhamme, J. (2018). How to write really good articles for premier academic journals. *Industrial Marketing Management, 68*(2), 202–209. https://doi.org/10.1016/j.indmarman.2017.11.014

Lee, B., Kazi, R. H., & Smith, G. (2013). SketchStory: Telling more engaging stories with data through freeform sketching. *IEEE Transactions on Visualization and Computer Graphics, 19*(12), 2416–2425.

Luzón, M. J., & Pérez-Llantada, C. (Eds.). (2019). *Science communication on the Internet: Old genres meet new genres*. John Benjamins. https://doi.org/10.1075/pbns.308

Marinov, R. (2020). Mapping the infotainment literature: Current trajectories and suggestions for future research. *The Communication Review, 23*(1), 1–28.

Martin, F., & Bolliger, D. U. (2018). Engagement matters: Student perceptions on the importance of engagement strategies in the online learning environment. *Online Learning Journal, 22*(1), 205–222.

Massarani, L., & Moreira, I. C. (2004). Popularisation of science: Historical perspectives and permanent dilemmas. *Quark, 32*, 75–79. https://raco.cat/index.php/Quark/article/view/55039/63356

Mersey, R. D., Malthouse, E. C., & Calder, B. J. (2010). Engagement with online media. *Journal of Media Business Studies, 7*(2), 39–56.

Niemann, P., Bittner, L., Schrögel, P., & Hauser, C. (2020). Science slams as edutainment: A reception study. *Media and Communication, 8*(1), 177–190. https://doi.org/10.17645/mac.v8i1.2459

Nikoloski, K. (2015). Charismatic leadership and power: Using the power of charisma for better leadership in the enterprises. *Journal of Process Management—New Technologies, 3*(2), 18–26.

Oliveira, A. W., Brown, A. O., Carroll, M. L., Blenkarn, E., Austin, B., & Bretzlaff, T. (2021). Developing undergraduate student oral science communication through video reflection. *International Journal of Science Education, Part B, 11*(2), 143–154. https://doi.org/10.1080/21548455.2021.1907630

Parsons, E. (1937). *The structure of social action* (2nd ed., 1968). Free Press.

Qiu, X., & Jiang, F. K. (2021). Stance and engagement in 3MT presentations: How students communicate disciplinary knowledge to a wide audience. *Journal of English for Academic Purposes, 51*, 100976. https://doi.org/10.1016/j.jeap.2021.100976

Reynolds, S. (2013). How to be more charismatic. Forbes. Retrieved March 18, 2022, from https://www.forbes.com/sites/siimonreynolds/2013/09/15/how-to-be-more-charismatic/?sh=d14b6594864d

Riesch, H. (2015). Why did the proton cross the road? Humour and science communication. *Public Understanding of Science, 24*(7), 768–775. https://doi.org/10.1177/0963662514546299

Riggio, R. E. (1998). Charisma. *Berkshire Bytes*, 1–9. Retrieved March 21, 2022, from https://www.berkshirepublishing.com/assets/pdf/Charisma_Byte.pdf

Rowley-Jolivet, E., & Carter-Thomas, S. (2019). Scholarly soundbites: Audiovisual innovations in digital science and their implications for genre evolution. In M. J. Luzón & C. Pérez-Llantada (Eds.), *Science communication on the Internet: Old genres meet new genres* (pp. 81–106). John Benjamins. https://doi.org/10.1075/pbns.308.05row

Ruiz-Madrid, N., & Fortanet-Gómez, I. (2015). A multimodal discourse analysis approach to humour in conference presentations: The case of autobiographic references. *Procedia—Social Behaviour Science, 173*, 246–251. https://doi.org/10.1016/j.sbspro.2015.02.060

Scalice, D., Dolci, W., Brochu, L., Merriman, T., Davis, H., Billings, L., & Voytek, M. A. (2019). FameLab USA: Improving science communication skills for early career scientists. *Astrobiology, 19*(4), 614–623. https://doi.org/10.1089/ast.2017.1809

Schaufeli, W. B. (2013). What is engagement? In C. Truss, K. Alfes, R. Delbridge, A. Shantz, & E. Soane (Eds.), *Employee engagement in theory and practice* (pp. 29–49). Routledge.

Shils, E. (1965). Charisma, order, and status. *American Sociological Review, 30*, 199–213.

Shils, E. (1968). Charisma. *International Encyclopedia of the Social Sciences, 2*, 386–390.

Soto-Sanfiel, M. T., Villegas-Simón, I., & Angulo-Brunet, A. (2022). Understanding the lack of science on TV: Network decision-makers' opinions towards televisual science. *Estudios sobre el Mensaje Periodístico, 28*(1), 227–239. https://doi.org/10.5209/esmp.72837

Stimm, M. (2020). *Science slam: ein Format der Wissenschaftskommunikation aus erwachsenenpädagogischer Perspektive*. Transcript. https://doi.org/10.1007/978-3-658-28861-7_10

Stuart, A. E. (2013). Engaging the audience: Developing presentation skills in science students. *Journal of Undergraduate Neuroscience Education, 12*(1), A4–A10.

Sun, H., & Linton, J. D. (2014). Structuring papers for success: Making your paper more like a high impact publication than a desk reject. *Technovation, 34*(10), 571–573.

Takahashi, B., & Tandoc, E. C., Jr. (2016). Media sources, credibility, and perceptions of science: Learning about how people learn about science. *Public Understanding of Science, 25*(6), 674–690.

Tubbs, R. S. (2019). Learning charisma. In M. Shoja, A. Arynchyna, M. Loukas, A. V. D'Antoni, S. M. Buerger, M. Karl, & R. S. Tubbs (Eds.), *A guide to the scientific career: Virtues, communication, research and academic writing* (pp. 103–107). Wiley Online Library. https://doi.org/10.1002/9781118907283.ch13

Valeiras-Jurado, J., & Bernad-Mechó, E. (2022). Modal density and coherence in science dissemination: Orchestrating multimodal ensembles in online TED talks and youtube science videos. *Journal of English for Academic Purposes, 58*, 101118. https://doi.org/10.1016/j.jeap.2022.101118

Wiseman, R. (2008). *Quirkology: The curious science of everyday lives*. Pan Macmillan.

Xia, S. A., & Hafner, C. A. (2021). Engaging the online audience in the digital era: A multimodal analysis of engagement strategies in TED talk videos. *Ibérica, 42*, 33–56.

Youknovsky, A., & Bowers, J. (2020). *Sell your research: Public speaking for scientists*. Springer.

Zarkadakis, G. (2010). FameLab: A talent competition for young scientists. *Science Communication, 32*(2), 281–287. https://doi.org/10.1177/1075547010368554

12

Research Visibility and Speaker Ethos: A Comparative Study of Researcher Identity in 3MT Presentations and Research Group Videos

Elizabeth Rowley-Jolivet and Shirley Carter-Thomas

1 Introduction

In today's increasingly competitive research world, researchers need to be visible and establishing an appropriate academic identity is vital. We propose to characterise the identities projected by speakers in two spoken academic genres situated at different stages in researchers' careers: three-minute thesis (3MT) presentations by doctoral students and research group videos (RGVs) by researchers in university laboratories.

E. Rowley-Jolivet (✉)
Université d'Orléans, Orléans, France
e-mail: elizabeth.jolivet@univ-orleans.fr

S. Carter-Thomas
IMT Business School, Evry, France
e-mail: Shirley.Thomas@imt-bs.eu

R. Plo-Alastrué, I. Corona (eds.), *Digital Scientific Communication*,
https://doi.org/10.1007/978-3-031-38207-9_12

We adopt a discursive, socially constructed view of identity (cf. Flowerdew & Wang, 2015), which is seen as fluid, multi-layered and dynamic, varying with the social context of communication, the person's membership of a particular culture or community, the audiences addressed and the particular facets of identity that are selected or foregrounded, positing that researchers make deliberate choices to project a particular identity or *ethos* (Amossy, 2010). While this view of identity can be traced back to Goffman's seminal study (Goffman, 1959) on selective self-presentation and identity as performance, it has acquired particular salience in the context of late modernity (Giddens, 1990), where the waning of tradition has weakened socially prescribed roles and accentuated the variety of identities that individuals can project. Furthermore, with the advent of Web 2.0 and the digitisation of the knowledge economy (Barbour & Marshall, 2012), the range of genres, semiotic resources and platforms through which academics can now perform various identities has been immeasurably extended. As pointed out by Darvin (2016), digital media provide multiple spaces where users can perform multiple identities. The shift towards the disintermediation of academic research has led scholars to present themselves and their work directly online, using the digital medium to create online personas, develop professional identities and claim a place both inside and outside their academic communities (Luzón & Pérez-Llantada, 2022).

Academic institutions, in particular universities, have been quick to seize the opportunities offered by digital media to promote their wares. In "the increasingly marketized and entrepreneurial higher education system world-wide [...] driven by an intense rivalry among institutions forever competing for resources and recognition" (Herman & Nicholas, 2019, p. 2), one of the most valuable assets in the scholarly marketplace is researchers' reputation and productivity, at all levels of academia from novice (doctoral students) to senior academics and prestigious research groups. Achieving visibility for this output—accomplished nowadays largely through an online presence—is therefore a prime concern for universities. This trend is illustrated by the two digital genres selected here, both of which are hosted on university platforms. At the same time, these developments have, however, raised several challenges for both researchers and research institutions: how to successfully manage and navigate between different genres, how to communicate esoteric research findings in an environment of context collapse (Marwick & Boyd, 2011), how to train researchers in the new communication skills required (Carter-Thomas & Rowley-Jolivet, 2020) and how to negotiate different research identities: professional, disciplinary, institutional and individual.

Among the different digital genres available for research communication, we focus on spoken academic communication, which is still relatively under-researched compared to written web-mediated documents. More specifically, we have chosen to study videos, as several studies have shown that online video is the preferred option to disseminate science (see e.g., Erviti & Stengler, 2016; León & Bourk, 2018). To explore whether video is equally suitable for the online dissemination of humanities research, our corpus comprises an equal number of videos in science, technology, engineering, maths and medicine (STEMM) and in the social sciences and humanities (SSH) disciplines. The two genres selected, however, 3MTs and RGVs, do not stand in the same relation to digital media. 3MTs are initially delivered to a live audience and the video recordings of the successful candidates (winners and finalists) are subsequently uploaded, without any digital post-processing, to the university's website or YouTube channel. They can therefore, like TED talks, be considered 'replicated' genres (Shepherd & Watters, 1998), with a small, primary audience (the live presentation) and a vast, secondary audience (YouTube viewers worldwide). 3MTs also have to follow strict rules (see Sect. 2). The RGVs, in contrast, are web-native genres, produced specifically for online diffusion, have no explicit conventions or rules to respect and can exploit the full range of multimodal resources. Consequently, while both the 3MTs and the RGVs in our corpus are digital in that they are internet-mediated, the means used by 3MT and RGV speakers to construct and project their identities are likely to differ greatly.

After presenting the corpus and method in Sect. 2, we then explore the semiotic resources available for identity construction in each genre, focusing first (Sect. 3) on non-verbal modes and then (Sect. 4) on the verbal mode through a linguistic analysis of the transcriptions.

2 Corpus and Methodology

2.1 Corpus

For this study, we used a previously collected corpus of 30 3MT presentations, half in STEMM and half in SSH (Carter-Thomas & Rowley-Jolivet, 2020), and collected a second corpus of 30 RGVs, again half in STEMM and half in SSH (see Table 12.1).

All the 3MT speakers were carrying out their doctoral research in an English-speaking country (Australia, Canada, United Kingdom, United States) and the majority were native English speakers. The talks were downloaded either from the respective university websites or from the universities' YouTube channels. In selecting from among the large number of 3MT talks available online, an additional criterion was that speakers had to be the winner or runner-up of their competition, in order to ensure that the talk was considered by the jury an excellent example of the genre. As our 3MT corpus was collected in 2011–2016, before the COVID-19 epidemic, the talks were delivered to a live audience, not pre-recorded in a studio and delivered to the camera (cf. the 3MT corpus of Beltrán-Palanques, this volume).

The aim of the 3MT contest, as defined on the University of Queensland website,[1] is to cultivate "students' academic, presentation, and research communication skills" and "their capacity to effectively explain their

Table 12.1 The corpus

	Number of 3MT talks	Average length of talks (in words)	Number of RGVs	Average length of RGVs (in words)
Science	15	494	15	503
SSH	15	464	15	569
Total	30	14 420	30	16080

[1] Source: https://threeminutethesis.uq.edu.au/participating-institutions, last consulted 22 April 2022. The University of Queensland (UQ) launched the competition in 2008 and continues to keep a fairly tight hold on it, with 3MT being a registered trademark of UQ. Any institutions wishing to hold a 3MT contest are supposed to request permission, use the 3MT brand on any materials, and abide by the rules laid down by UQ.

research in three minutes, in a language appropriate to a non-specialist audience". As this description makes clear, the focus is on communication skills (rather than on research skills) and on targeting a non-specialist audience. Competitors must abide by strict rules: those who overstep the 3-minute time limit are automatically disqualified, speakers are allowed only one static slide and no other props or audiovisual files. Competitors are thus deprived of one of the main warrants (Toulmin, 1958) used in academic argumentation to support and provide evidence for the claims, namely visual evidence. The most important defining feature, however, is that it is a competition. Participants have to give a one-off performance. There is no question time at the end to explain or clarify matters for the audience as for a conference presentation, and the decision of the adjudicators is final.

The RGVs, while not part of an explicit contest like 3MTs, nonetheless participate of the same promotional and competitive spirit, vying for attention in the online environment. In Luzón's words, "online videos are exploited by academics to disseminate information about science to the public and to promote their work and findings" (Luzón, 2019, p. 168). As they are delivered by experienced researchers, however, the formative purpose of 3MTs—cultivating students' communication skills—is not an issue.[2] The RGVs were downloaded from the YouTube channels of the top 10 research universities in the worldwide ranking system so as to guarantee a homogenous quality of the research presented. To ensure temporal comparability with the 3MT presentations, we selected short videos that lasted between 2 and a half and 4 minutes, and that had been uploaded in the last 5 years (2015–2020). In SSH fields, this proved difficult as many of the available videos were much longer than our 2½–4 minute time limit, indicating either that the use of these short-form 'scholarly soundbites', which have become popular in STEMM fields to disseminate research to a wide audience (Rowley-Jolivet & Carter-Thomas, 2019), has not yet caught on in SSH, or that some areas of SSH research do not lend themselves well to very brief presentations. Two of

[2] Space is lacking here to give a detailed analysis of the differences in rhetorical structure between the two sub-corpora. See Sect. 4.1, Positioning towards topic and audience, however, for a brief outline of this aspect.

the 15 SSH RGVs were therefore collected from other English-speaking universities (Edinburgh and Queensland). We selected only videos that presented a university research group, and that were delivered by the researchers themselves, that is, we excluded interview-based videos involving a journalist, voice-off commentaries, videos that presented a department, a course of study or the university in general.

This gave us two sub-corpora that were comparable in disciplinary scope, length and time period. All 60 recordings were transcribed by the authors. The resulting oral corpus, after transcription of the videos, comes to 30,500 words.

2.2 Method

For the analysis of the semiotic resources available, the initial methodological decision to make was between a micro-analysis, that is, full multimodal analysis of the various modes simultaneously co-deployed, or a macro-analysis of the whole corpus. The two methods are complementary, each having its advantages and drawbacks. Full multimodal analysis enables the interplay between multiple modes to be captured in great detail as the talk unfolds, but is limited to extremely brief extracts (a few tens of seconds of video, e.g., Xia & Hafner, 2021; Harrison, 2021), cannot inform about the overall features of the genre and produces results that are highly speaker-dependent. A macro-analysis, in contrast, is less suitable for exploring modal combinations or multimodal ensembles, but allows quantitative and qualitative analyses of a much larger amount of data that reveal recurrent features and dominant characteristics. As our aim was to compare identity construction in two genres, 3MTs and RGVs, we opted for the latter method in order to propose an overview of the whole corpus (totalling 3 hours of video recordings). Using a corpus-driven approach based on close observation of the data, we therefore analysed the non-verbal and verbal resources separately.

Among the non-verbal resources, a major distinction relevant for our corpus is that between embodied modes, defined here as modes performed with the speaker's body, such as vocal effects, posture or movement, gesture, gaze and facial expression (Wachsmuth et al., 2008) and filmic modes incorporated during postproduction and editing by the professional media service that produced the video, such as stills, moving images, text, sound effects, animations and graphics. We additionally drew on Luzón's study of online videos by scientific research groups, in particular the semiotic resources used to establish the group's authority and credibility (Luzón, 2019).

For the verbal resources, using move structure (Swales, 1990) we firstly analysed the type of content selected by the speakers before focusing on various interactional resources including the use of first person pronouns that can contribute to creating a particular identity and rapport with the audience. In this second stage, AntConc 3.4.0[3] was used to study the initial quantitative distribution of the linguistic features.

3 Non-verbal Identity Construction

We follow Kress et al. (2001) in assuming the 'functional specialisation' of different modes, that is, "there are some things that some modes have been developed to do better than others. The meaning-making potentials of the resources of the visual, actional and linguistic modes each perform a special and differently significant role" (Kress et al., 2001, p. 16). Given the pronounced differences in the modal resources available in 3MTs and RGVs to project their identities, each genre will therefore be dealt with separately in this section.

3.1 Non-verbal Identity Construction in 3MTs

In 3MTs, almost the only non-verbal resources available for identity construction are embodied modes—vocal effects, posture or movement, gesture, gaze and facial expression. Visual communication *stricto sensu* is limited to the single static slide that is projected as a backdrop to the talk and the speaker is alone on a bare stage in front of a live audience: as in a sketch in a one-man or one-woman show, speakers have three minutes in

[3] https://www.laurenceanthony.net/software/antconc/releases/AntConc340/.

which to capture and hold audience attention, without any props or audiovisual aids (moving images, music or sound effects, other special effects, etc.) and thus rely solely on their individual, physical presence. In this genre, the notion of identity proposed by Benwell and Stokoe can indeed be understood not only as a construction but as a public performance in the theatrical sense of the word:

> identity is a public phenomenon, a performance or construction that is interpreted by other people. This construction takes place in discourse and other social and embodied conduct, such as how we move, where we are, what we wear, how we talk and so on. (Benwell & Stokoe, 2006, p. 3)

All five embodied modes are used by the winning contestants to interact with the audience and perform their identity. Speakers bring their doctoral research to life by exploiting the expressive possibilities of vocal effects, adopting a lively and energetic delivery, placing vocal stress on key terms to help comprehension or express conviction, using interrogative and exclamatory intonation both to structure their presentation and to attract attention:

(1) think of the emotions aroused in the first two scenarios. SADness, desPAIR, LOVE—these are STRONG emotions. (3MT, SSH)

(2) And you're all looking at me and thinking one thing—Ouf! he's hot. (3MT, STEMM)

Proxemics are another embodied mode available to 3MT speakers: as they are unencumbered by notes, a rostrum or the need to manipulate slides, they are free to move around on the stage and fill the space, adopting an open body position towards the audience and alternating standing, pacing and other footwork to dynamise the talk and create variety, rather than remaining static as is usually the case during the delivery of conference presentations (see Fig. 12.1). Gaze direction is principally 'mutual gaze', that is, audience-directed, throughout the talk, except when speakers shift their gaze to the gesture they are making in order to focus mutual attention on it and give it salience.

Fig. 12.1 Body posture accompanying the speaker's words "Would you believe me if I told you this was my brain on drugs?" in a 3MT talk (STEMM). Note: All the illustrations are original, specially commissioned drawings by Hai-Hsin Huang based on screenshots from the video recordings

Gesture is an abundantly used mode, constantly co-deployed with verbal language in the talks. Although the absence of any visual support apart from their single slide could prove a severe handicap in many research fields, particularly STEMM, where visualising the phenomenon or process is an integral part of communication, speakers compensate for this lack as much as possible by exploiting the explanatory and expressive properties of gesture. All four categories of gesture defined by McNeill (1992), namely iconics, metaphorics, deictics and beats, are used. Iconic gestures (Fig. 12.2a) serve mainly to explicate technical content, creating proximity with the lay audience by making the research accessible: in STEMM, for instance, precise hand, finger and arm movements and the different force and amplitude with which the gestures are performed enable viewers to follow the development of processes in 3D and to grasp accompanying abstractions such as speed, strength, resistance and so on. Metaphoric gestures are liberally used by both SSH and STEMM speakers either to convey abstract notions (such as rejection in Fig. 12.2b) or to symbolise the giving of information and interpersonal interaction with the audience (e.g., palm up open hand), while deictics (pointing gestures) and beats (rhythmic up-down hand movements), generally co-deployed

Fig. 12.2 (**a**) Iconic gesture accompanying the words "Now if we heat the device from underneath" (3MT, STEMM), (**b**) Metaphoric gesture accompanying the words "Protestants came to REJECT the majority of these interpretations" (3MT, SSH), (**c**) Deictic gesture accompanying the words "THIS is exciting!" (3MT, SSH), and (**d**) Beat gesture accompanying the words "as mum always said, 'there's MORE [*beat*] to LIFE [*beat*] than DRUGS [*beat*]'" (3MT, STEMM) in 3MTs

with vocal effects, express their enthusiasm and involvement in their work (Fig. 12.2c, d).

Lastly, facial expression: here also, 3MT speakers exploit the expressive possibilities of the mode, smiling, raising their eyebrows or adopting other facial expressions to suit their words, while nonetheless avoiding clowning or overacting (see e.g., Figs. 12.1 and 12.2b, c). Some speakers even adapt their physical appearance (hairstyle, clothing) to suit their topic, visually identifying their whole body with the theme presented. For instance, the speaker in Fig. 12.2a, who gave his talk on heat (thermal electronics), wore a T-shirt showing red flames, while the speaker in Fig. 12.1, whose talk concerned the delivery of drugs to the brain, chose a hairstyle in psychedelic colours. The overall impression conveyed by combining these different embodied resources is that of a young, enthusiastic and dynamic person, capable of a certain theatricality to convey their work in an entertaining manner, friendly and open to communication with others and capable of modal flexibility and adaptability to the context.

Their institutional identity is of course a given of the communicative context; as the contest is organised by the university, the name and logo of the institution are often displayed on the stage and always constitute the opening screen of the video recording posted on YouTube. However, the main thrust of the presentations is audience interaction rather than foregrounding the academic credentials of these novice researchers. This is also illustrated by the types of slide shown. The single static slides allowed do not project an esoteric academic identity by displaying the sort of figure characteristic of research articles; rather, they provide the audience with an evocative and striking image to interest or intrigue them, using saturated colours, 'demand' images and emotively striking pictures. One speaker in SSH, for instance, whose talk is on emotions and learning, shows three rather grim masks each representing an emotion such as fear, frustration or sadness (see Rowley-Jolivet & Carter-Thomas, 2020, for some examples of slides).

3.2 Non-verbal Identity Construction in RGVs

In the RGVs, the situation is very different, as in addition to the embodied modes performed by the researchers in the video, they can also exploit all the filmic modes of online videos. These multimodal resources are abundantly used in many popular science videos on the web (see e.g., Muñoz Morcillo et al., 2016), which are often very fast-paced, dramatic and visually impactful, with a high technical complexity of montage. The RGVs in our corpus are considerably more restrained in their use of filmic modes but are nonetheless carefully crafted to project a certain identity (or identities) of the research group and its members. As in the linguistic analysis of speaker roles (see Sect. 4), several different identities can be distinguished: institutional, researcher and personal. Table 12.2 lists the non-verbal resources observed[4] and, where relevant, the number of videos in which they occur.

Our findings confirm Luzón's (2019) that RGVs firmly establish the institutional identity of the researchers. Their academic credentials (full academic titles and affiliations) are always printed on screen, the name (and also often the crest) of the university features in all 30 RGVs and we see many shots of the university buildings where they work. We also get occasional glimpses of their teacherly identity through shots of the researcher lecturing to students or giving seminars.

The main focus of both the verbal commentary and the video footage, however, is their researcher identity, depicted in three main environments: the lab, their university office and out in the field. Not only do they talk about their research while seated at their desk, but the video medium allows aspects of this identity to be communicated that are difficult to convey in other modes: we see them actually doing research—conducting experiments in the labs, working on their computers, out on field expeditions or working with pupils and teachers in schools and other external collaborators. The RGVs give a hands-on image of research as an activity that makes a valuable contribution to the general community and to societal issues, while the researchers' peer-directed writerly identity (producing research articles), on which their credibility in their academic discourse communities is established, is almost totally absent: we found only one reference each to a research article in STEMM and SSH. Their expertise is, however, underlined, as visually attested by the sophisticated lab and field equipment in STEMM (Fig. 12.3a) and by the impressive shelves of books in SSH (Fig. 12.3b). Unlike in many scholarly blogs, on the other hand, glimpses of the researcher's private life and

[4] As we were not doing a micro-analysis using multimodal analysis software, we did not take the more technical filmic modes such as cuts and types of camera shots or angles into consideration.

Table 12.2 Non-verbal resources for identity projection in the RGVs

	STEMM	SSH	Examples
FILMIC MODES			
Institutional Identity			
1 Shots of university buildings and research centres (outside and inside)	4	8	All Souls College, Oxford Centre for Digital Humanities, Princeton University lecture halls, libraries, seminar rooms, staff rooms
2 Academic status and affiliation	15	15	Researcher's name, title (Prof., Dr, post-doc), position occupied, name of university and/or research laboratory
3 Name (& crest/logo) of the university	15	15	Displayed on the opening and/or closing screens, sometimes on screen throughout the video
4 Teaching role	1	1	Speakers shown lecturing or giving seminars
Researcher Identity			
5 research in the lab	14	1	STEMM: Animations, schematics and video footage of methods and processes to clarify and illustrate the research; high-tech lab equipment shown and operated; teamwork SSH: The sole lab environment is in archaeology (equipment for weighing and measuring bones)
6 research in their office	11	9	Researchers surrounded by their computers, shelves of books

(*continued*)

Table 12.2 (continued)

	STEMM	SSH	Examples
7 research in the field	6	7	STEMM: Antarctica (counting penguins), on boat (sea-floor expedition), collaborating with colleagues in India, diving in the Maldives (tracking Manta rays), crop-growing SSH: collaborating with teachers in schools, with museum curators, with actors
Personal Identity			
8 Personal life	2	0	Researcher with her baby daughter in the street Researcher with family photos
Special effects			
9 Sound effects	15	13	Very discreet background music in 28 videos No other special sound effects
10 Visual effects	0	1	No special filmic effects apart from fast motion in 1 SSH video
EMBODIED MODES			
11 Dress codes	STEMM: Casual everyday dress, appropriate to the working environment (lab coat, T-shirt and jeans, anoraks in the field etc.) SSH: Female researchers dress quite smartly (make-up, pearls…)		
12 Gaze, vocal effects, gesture, movement	Not a stage performance: natural delivery (no histrionics, no lecturing style) and movements or gestures as they work. Gaze directed at the camera or towards the task in hand or colleagues		

identities are extremely rare: in our corpus, we found only a couple of allusions, both of which serve to underscore the researcher's personal motivation for his/her work.

There are some differences, however, between the researcher identities of STEMM and SSH researchers portrayed in the videos, which appear

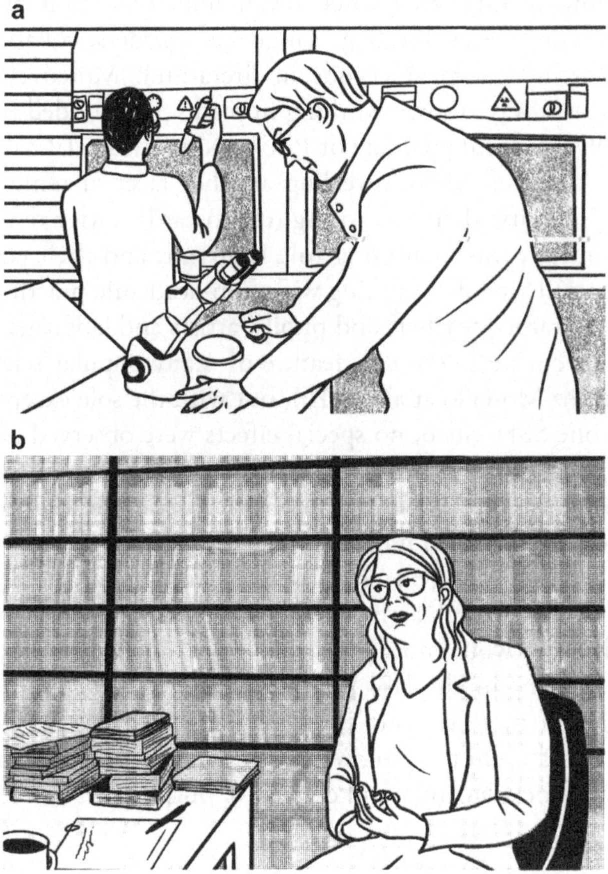

Fig. 12.3 Researcher identity projection in RGVs. (**a**) Science researcher in his lab (RGV, STEMM), (**b**) Humanities researcher in her office (RGV, SSH)

to be related to the different 'epistemic cultures' (Knorr-Cetina, 1999) of the two fields. As shown in Table 12.2 and Fig. 12.3a, in STEMM, the lab is the main locus of researchers' activity. One of the main advantages of filmic modes is that they enable lab research to be made more concrete and easily understandable for the non-specialist audience by visualising it: animations, schematics and video footage of methods and processes are abundantly used in STEMM RGVs to clarify and illustrate the research, as also observed by Luzón (2019). This affordance is much less relevant for many SSH fields, where research focuses on intangible topics that are difficult to visualise (e.g., language disorders, collaborative linguistic research, ecological visions in literature). Moreover, STEMM researchers are always shown working in teams, surrounded by lab technicians, colleagues and graduate or PhD students (Fig. 12.3a), several of whom speak in the videos, revealing another facet of senior STEMM researchers' identity: their mentoring role towards novice researchers. In SSH, in contrast, this mentoring role is absent, and their collaborative projects generally involve working with non-academic practitioners such as museum curators, teachers and pupils, artists and language therapists.

Special effects are a common feature of many popular science videos online (Muñoz Morcillo et al., 2016), but with the sole exception of fast motion in one SSH video, no special effects were observed in our RGV corpus. Similarly, sound effects are very discreet, consisting solely of quiet background music in 93% of the videos. A similarly restrained style was observed in the embodied modes: unlike 3MT presenters, researchers in the RGVs are not putting on a stage performance but are portrayed going about their usual research activities. They speak simply and naturally and are dressed as they would normally be for work (anoraks in the Antarctica, lab coats or jeans in the lab, jackets and open-necked shirts in their offices), though more casually in STEMM than in SSH. Gestures are much less called upon than in 3MTs as moving and static images can be used instead for explanatory and expressive purposes.

4 Verbal Expressions of Identity

Both the RGVs and 3MTs target non-specialist audiences who may have little or no knowledge of the precise topic. The subsequent recontextualisation that is needed to make the content accessible requires changes in the roles adopted by the speakers and the degree of proximity or authority they wish to communicate. We will examine the verbal means speakers use to establish this identity in the two genres under two main headings: (a) positioning towards topic and audience and (b) establishing a persona.

4.1 Positioning Towards Topic and Audience

For their talks to be accessible, speakers need to decide which aspects to focus on or on the contrary to downplay, whilst considering the constraints and affordances of the genres they are working in. In a previous study on the move structure of 3MTs (Carter-Thomas & Rowley-Jolivet, 2020), we showed how certain categories of content are highlighted, namely: (a) the orientation, (b) the rationale for the study and its purpose, (c) its implications for the future and (d) a termination move or coda. Other categories, such as methodology and theory, are omitted or only very shallowly evoked. Results were not always mentioned, probably because some contestants were only in the second year of PhD research and thus did not yet have any results to present. Using similar criteria, we also identified the preferred move categories in the RGVs, where a rather similar profile emerged, with three main aspects foregrounded: (a) the research rationale—what are we doing and why is it important? (b) the results: what has been achieved? and (c) the applications of the research or its usefulness.

The orientation move is a prominent feature of all the 3MT presentations. To capture the audience's attention from the start, several 3MTs begin in quite a theatrical fashion with a verbal 'hook'—a direct question addressed to the audience, a moving personal story, a striking fact or event or an imaginary scenario. Scenarios, which are specific to 3MTs (Carter-Thomas & Rowley-Jolivet, 2020), evoke a possible but imaginary situation that allows the speaker to explain or introduce a complex event and are very dramatic:

> (3) Picture a monster that can attack at any time, that can change itself to evade the weapons used against it and can come back to life from the dead. [...] It's killed Steve Jobs, Patrick Swayze and when I was fourteen my mother. [....] the monster is cancer. (3MT, STEMM)

Other strategies used by 3MT speakers to capture the audience's attention include the use of personal stories (some presenters recount very intimate details of their lives, such as the death of a mother as in (3) or the learning difficulties of a son) as well as what we have previously termed 'street cred' (ibid.). Here the aim is to create a common ground based on shared cultural values and interests, rather than scientific ones by referring for example, to films or TV series:

(4) Contrary to what you see on CSI, it's not computers that match prints, it's humans. (3MT SSH)

Speakers present themselves not only as budding academics but as real, approachable people with interests and concerns that many in the audience can identify with. Their authority is not based on the same criteria as those used by examiners at the thesis viva, namely scientific know-how. Instead, 3MT speakers rely more on non-technical arguments, using personalisation and interactional strategies to create solidarity.

The opening sequences in the RGVs are more restrained, with no theatricality and only a couple of personal stories. Instead, speakers emphasise the newsworthiness of their research by presenting its aim and importance (5) or fronting the results (6), adopting the 'inverted pyramid' structure of news reports:

(5) The Crop Science centre is going to make a huge difference. It amplifies our impact. It puts us I think in a very very strong global position. (RGV STEMM)

(6) Bringing archives to light and into the classroom and making archives available at students' fingertips that would previously be really difficult to find or to access—it's really revolutionised the study of a lot of texts from the past. (RGV SSH)

One of the main aims in producing RGVs is to enhance the visibility of the research by showing the group's achievements (Luzón, 2018). It is, however, the social applications and/or practical usefulness of the research, not its theoretical import that is emphasised:

(7) we hope that in the near future [this test] will be used to triage patients with active TB and get them the treatment they need, potentially saving millions of lives. (RGV STEMM)

Despite the fact that the videos came from nine different universities and covered a wide range of disciplines, and that there are no generic conventions, we found a strong homogeneity across the board in their general organisation and style. There appear to be some tacit constraints at work that govern how research is communicated. For instance, the fact that all our RGVs are posted on university websites and endorsed by the respective universities means that the university's reputation is at stake, as are the researchers' own digital reputations. Expertise and seriousness are therefore what are underlined. This aspect is dealt with in greater detail in Sect. 4.2.

This expertise is not, however, linked to the defence of a claim, as is the case in genres such as the research article. We found only two references to the literature in the RGV sub-corpus. The RGVs seem to take place in a rhetorical vacuum: there is no controversy, no discussion of competing theories and claims and no academic argument. The RGVs give a non-confrontational, non-agonistic picture of academic research. Although the theoretical background move is not a strong feature of either genre, in the 3MTs theory plays a slightly more important role. Despite the quasi-absence of literature references, speakers quite often provide general background knowledge of their topic, as remarked on below in the discussion of the plural pronoun *we*.

A final remark in relation to content organisation concerns the closing move, where there is a marked difference in speaker positioning in the two genres. In 3MTs, the speakers usually round off their presentation by picking up audience contact again and making an informal remark:

> (8) So, next time you see some alien markings on the ground you now know who you're going to call. Thank you. (3MT SSH)

In the RGVs, on the other hand, the verbal commentary ends in practically all cases with an emphasis on the social significance of the research, although some also add a striking <u>visual</u> termination move such as the spectacular birth of a Manta Ray in captivity or a hugely magnified image of a pink blob that constantly changes its shape, showing cell division taking place in the strange species *Tricoplax adhaerens*. Given that the web audience is in theory infinite and potentially diversified, it is difficult

Table 12.3 Questions in the 3MTs and RGVs

		Science	SSH	Total
3MT talks	Number of questions	35	41	76
	Per 10,000 words	47.2	58.5	52.7
	Coverage	11/15	12/15	23/30
RGVs	Number of questions	7	8	15
	Per 10,000 words	9	9	9
	Coverage	4/15	5/15	9/30

for speakers to rely on common knowledge or interests, whether cultural or scientific. The relationship with the audience is one that is based more on trust and respect, rather than on solidarity.

The difference in speaker positioning is probably linked to the presence of a live audience in 3MTs, where engagement strategies such as questions and second person pronouns are also more frequently used. Table 12.3 shows that 3MTs in both SSH and Sciences use six times as many questions as RGV speakers do (52.7 vs. 9 per 10,000 words).

Some 3MT speakers open their talk with a question addressed directly in the second person to the audience, involving them in a pseudo-dialogue:

(9) Can you remember learning to write at school? (3MT SSH)

Question-and-answer pairs are another feature. 3MT speakers show their awareness of the audience's needs by foreseeing their queries and forestalling their reservations:

(10) So, you might think why on earth is she talking about this, why is she going round in circles? Well actually, this can prove to be quite a helpful reaction. (3MT STEMM)

In the RGVs, questions are not only infrequent, but when present, concern mainly the research questions addressed, rather than any explicit interpersonal concern:

(11) In the new world you sequence the genome of the bacterium and from that you derive all the information you need to answer all of the questions

you want to ask: what species is it? which other TB samples is it related to? which drugs is it resistant to and which will kill it? (RGV STEMM)

Proximity with the 3MT audience is also created through the use of second person pronouns (cf. Hyland, 2001b). To compare the use of *you* in the two genres, we distinguished between two main roles: generic *you* and participant *you*. (Other less frequent uses were *you* as a discourse marker, e.g., *as you know*, and the final *thank you* in the 3MT presentations.) As Table 12.4 shows, in the 3MTs, *you* is mainly used in a participant role. In (12), for instance, the presenter appeals directly to the shared knowledge and habits of the audience as smartphone users:

(12) And any of you with a smartphone which I presume is most of you will know that lithium ion batteries aren't very good at powering it for a long period of time. (3MT STEMM)

In the RGVs in contrast, *you* is practically only used with its generic and encompassing value, that is, as an alternative to the more formal *one*:

(13) In order to detect a bacterial infection, you need to culture the bacteria. (RGV STEMM)

The sole participant *you* is in a video where the researchers are seen bidding goodbye to the penguins they are observing:

(14) Bye guys! see you next week! (RGV STEMM)

Table 12.4 Functions of *you* in the RGVs and 3MTs

Function	RGV#	3MT
Discourse particle	9	3
Thank you	0	23
Other	5	3
Participant	**1**	**97**
Generic	**120**	**59**
TOTAL	135	185

4.2 Establishing a Persona

In 3MT presentations, the speaker's personality plays an important role. Presenters try to avoid anything likely to bore their listeners and the tone adopted is light and chatty, with many evaluative adjectives and a fairly informal register of language:

(15) Well I can see that you're absolutely fascinated [*laughter*] and you want to know why. (3MT, STEMM)

(16) Now this is brilliant. (3MT, STEMM)

Above all the aim is to be entertaining.

RGV speakers, while also keen to appear likeable and project a friendly persona, need to appear competent and worthy of the trust the sponsors and university administrations have placed in them. As a result, we find little of the rather bubbly enthusiastic language that is characteristic of the 3MT speakers. There is no slang or jargon and delivery remains neutral and calm. Drama and humour are not used and very few of the videos begin with a personal presentation of the speaker or speakers, as their names and academic credentials are already given on the screen.

One of the most explicit linguistic signs of the personal involvement of speakers is in the use of the first-person pronoun, corresponding to a conscious choice to portray a particular stance and identity. In both RGVs and 3MTs first-person subject pronouns are globally frequent, in comparison[5] for example with more impersonal written genres, such as the thesis or RA, with over 25 occurrences per 1000 words overall (Table 12.5). Both 3MT and RGV speakers take responsibility for their research actions and express their opinions congruently. However, *we* largely predominates in the RGVs (300 occurrences) and *I* is slightly more frequent in the 3MTs. This is unsurprising as 3MTs are presented by one speaker, whereas there are often several speakers in RGVs. Moreover, 3MTs are initially delivered to a live audience, while RGVs are

[5] Hyland (2001a), for example, in a cross-disciplinary study covering all self-mention forms in RAs found only five cases per 1000 words.

Table 12.5 Occurrences of *I* and *we* in RGVs and 3MTs

	RGVs		3MTs	
	I	*we*	*I*	*we*
STEMM	40	149	83	107
SSH	73	151	114	77
All	113	300	197	184
Frequency per 1000 words	7	18.6	13.6	12.8
	25.6		26.4	

designed for web diffusion from the outset. As these factors influence how the speaker identities are constructed and perceived, we analysed each pronoun separately.

Identities Performed by *I*

We took as our starting point the idea that (digital) identities can be multiple and that individual speakers can move between different identities, performing different roles during their presentations. Building on past typologies (Hyland, 2002; Rowley-Jolivet & Carter-Thomas, 2005; Fløttum et al., 2006) we distinguished between:

(a) Personal identity (or ID), where it is the speaker's private identity that is underlined
(b) Researcher/Professional ID, where the professional role of the speaker is underlined
(c) Speaker ID, where the presenter intervenes in a meta-discoursal or interactional capacity

Figure 12.4 shows that in 3MTs, a large proportion of *I* are used with a researcher ID to describe and promote the individual thesis research:

> (17) So, I found the cells, now I need to find the scaffold: polymers. (3MT STEMM)

This contrasts with RGVs where exclusive *we* predominates over *I* (see below). In both 3MTs and RGVs, there is a tendency for SSH speakers to

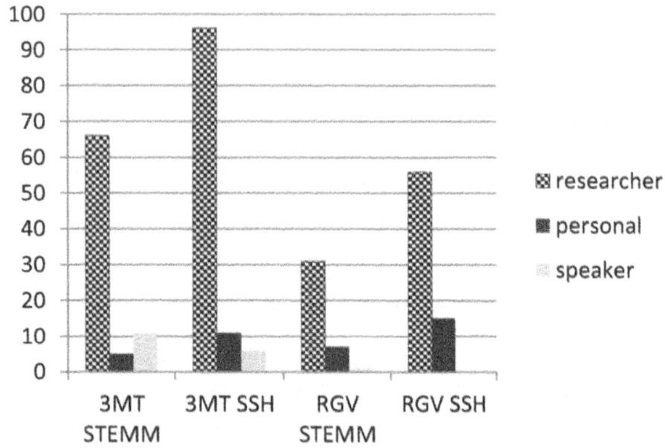

Fig. 12.4 Identities performed by *I* (in number of occurrences)

use *I* more than STEMM researchers, indicating that research at both novice and senior levels is also often conducted individually, not within a research group, in SSH. This was also very apparent in the SSH videos where senior and junior researchers are not seen working together, unlike in STEMM fields.

The pronoun *I* is also used by RGV and 3MT speakers to project a very personal identity, as in (18) and (19), where the speakers bring in an aspect of their personal lives as motivation for their research and where (19) explicitly underlines the multiple roles she performs in addition to that of PhD researcher:

(18) I grew up in an almost exclusively black church [...] And within that church there was imagery of a white male god that always just stood out to me. (RGV, SSH)

(19) I'm a teacher myself and a mother and an academic and through these various roles I've spent a lot of time in schools. (3MT, SSH)

The third category identified, Speaker ID, is only a feature of 3MTs, and is directly related to the live communicative context:

(20) By the time I've finished this three-minute talk, someone in the UK will have a stroke. (3MT STEMM)

Identities Performed by *We*

Some of these same tendencies can be observed when studying the use of *we*, where three values were distinguished:

(a) Exclusive *we*: The voice of the researchers as a group, constructing their identities as experts
(b) Inclusive *we*: Aligning with viewers and/or the general public, concerning shared knowledge and societal concerns
(c) Research community *we*: Aligning with the general 'state of the art' and summing up current research

As Fig. 12.5 shows, in RGVs the exclusive value of *we* largely predominates. There is on the contrary very little reference to shared public knowledge, with a low occurrence of inclusive *we* and an almost total absence of research community *we*. As noted above, there is very little mention of theory and/or reference to the work of other researchers in the field. The primary aim of these videos seems to be clearly to promote the work of the specific research group.

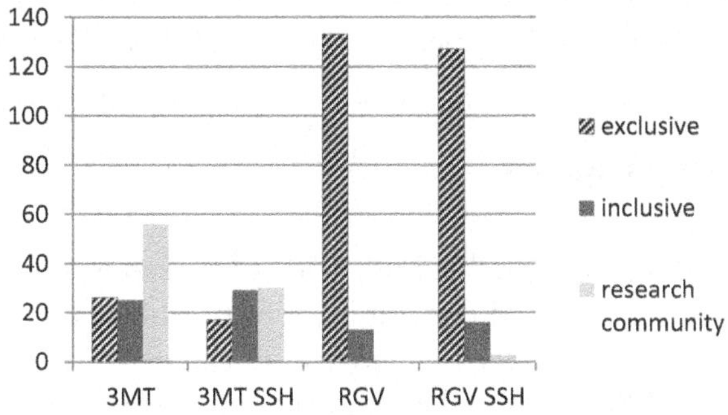

Fig. 12.5 Identities performed by *we* (in number of occurrences)

(21) What my group does, we apply an electric potential across this channel and we increase it slowly in time and what we find is that bacteria are immobilized. (RGV, STEMM)

A marked feature of 3MTs is the recourse to research community *we* to sum up and present background knowledge of the field. As novice researchers, the majority of 3MT speakers do not yet have any past research experience on which to draw. They are also not necessarily full members of a lab or group and as 'peripheral participants' (Lave & Wenger, 1991) maybe feel they aren't authorised to speak in the lab's name with an exclusive *we*. On the other hand, their PhD work involves a lot of reading, and summarising the literature and reporting this broader scientific knowledge, thus, probably looms large in their daily preoccupations.

(22) And using machine learning techniques we can do these sorts of analyses over millions of words in hundreds of languages. (3MT, SSH)

There is also a higher occurrence of inclusive *we* in 3MTs as the speaker tries to draw the live audience into the account of the research:

(23) During the nine months of development before birth we are lucky that we have our mothers there to protect us. (3MT, STEMM)

5 Conclusion

In this study of two digital genres, both addressed to non-specialist audiences, we have compared researcher identity along three dimensions: the level of researcher expertise (novice researchers in 3MTs vs. senior researchers in RGVs), disciplinary area (STEMM vs. SSH) and the different verbal and non-verbal affordances for identity projection. Results show, first, that in both genres, the researchers' credibility or *ethos* is grounded on different criteria to those conventionally expected in traditional written research genres. Instead of theoretical knowledge, detailed methodology and academic controversy, speakers use non-technical arguments such as social applications and practical outcomes to justify their work, in a concern for proximity with the lay audiences.

The comparison also reveals, however, numerous differences in the identities projected in the two genres, with the level of researcher expertise (novice vs senior) emerging as a strong factor of differentiation. In 3MT presentations, the entertainment value is paramount. Speakers put on a stage performance, creating the maximum impact through humour, moving personal details, expressive body language and vocal effects and theatrical opening sequences containing scenarios and stories. They use various strategies to interact and create solidarity with the live audience, such as participant *you*, direct questions or the creation of common ground through 'street cred' and a chatty register. Senior researchers, in contrast, behave as they would normally do in their working environment, speaking in natural but measured tones, working calmly in their labs or offices, with no recourse to drama, humour or scenarios. The image conveyed is of a likeable and approachable person but also someone who is both dedicated and highly competent.

We have also observed some disciplinary differences in the identities performed in STEMM and SSH. In both 3MTs and RGVs, the epistemic culture of SSH, at both novice and senior levels, appears to be more individualistic, while that of STEMM is collective, organised around the lab team. This can be observed in the tendency for SSH speakers to use *I* more than STEMM researchers and is also very apparent in the SSH RGVs where senior and junior researchers are not seen working together, unlike in STEMM where graduate and postgraduate students feature frequently in sequences shot in the lab, highlighting the strong mentoring role of senior STEMM researchers. SSH RGVs also make very sparse use of filmic modes such as animations, schematics and video footage of methods and processes, which appear to be much less relevant in SSH where research focuses on intangible topics that are difficult to visualise. This raises the question of whether short-form digital videos are in fact the most appropriate medium for the dissemination of SSH research, unlike in STEMM.

The digital affordances for identity projection in the two genres are also very different. 3MTs, like TED talks, are both an offline (live presentation) and online (digitally mediated) genre, in which almost the only non-verbal resources available to speakers for identity construction are embodied modes, underscoring the performance aspect of these talks, whereas RGVs can call on a wide range of filmic modes in addition to embodied expression. These filmic resources are mobilised to show the social usefulness and participatory nature of their research as well as to clarify the technical content for the audience, and allow them to project various facets of their identities—institutional, teacherly and researcher. RGVs clearly avoid using, however, any filmic modes that might appear gimmicky and detract from the professional credibility of the research group.

One can wonder, however, to what extent the identities performed/projected by researchers in 3MTs and RGVs are dictated by external factors beyond their control. In 3MTs, the very strict rules of the contest, the judging criteria imposed and the prior training received by contestants tend to create a kind of template for the performance, leading to the replication of certain communicative strategies, whoever the speaker may be and whatever the research topic. In RGVs, one cannot exclude the possible influence of two external factors that may have a considerable impact on identity projection: on the one hand, the media logics of the video production team, which are not research-related, and, on the other, the institution's agenda—indeed, some studies conducted from a critical perspective (e.g., Thoms & Thelwall, 2005) have claimed that the institution's use of digital media disempowers the researchers themselves, in that the identities of the individual are ultimately lost to the governmentality of the university.

Acknowledgements We warmly thank Hai-Hsin Huang for producing the original drawings from screenshots of the videos.

References

Amossy, R. (2010). *La présentation de soi. Ethos et identité verbale*. PUF.

Barbour, K., & Marshall, D. (2012). The academic online: Constructing persona through the World Wide Web. *First Monday, 17*(9).

Benwell, B., & Stokoe, E. (2006). *Discourse and identity*. Edinburgh University Press.

Carter-Thomas, S., & Rowley-Jolivet, E. (2020). Three minute thesis presentations: Recontextualisation strategies in doctoral research. *Journal of English for Academic Purposes, 48*, 100897.

Darvin, R. (2016). Language and identity in the digital age. In *The Routledge handbook of language and identity* (pp. 523–540). Routledge.

Erviti, M. C., & Stengler, E. (2016). Online science videos: An exploratory study with major professional content providers in the United Kingdom. *Journal of Science Communication, 15*(6), 1–15.

Fløttum, K., Dahl, T., & Kinn, T. (2006). *Academic voices: Across languages and disciplines*. John Benjamins.

Flowerdew, J., & Wang, S. H. (2015). Identity in academic discourse. *Annual Review of Applied Linguistics, 35*, 81–99. https://doi.org/10.1017/S026719051400021X

Giddens, A. (1990). *The consequences of modernity*. Cambridge Polity Press.

Goffman, E. (1959). *The presentation of self in everyday life*. Doubleday.

Harrison, S. (2021). Showing as sense-making in oral presentations: The speech-gesture-slide interplay in TED talks by Professor Brian Cox. *Journal of English for Academic Purposes, 53*.

Herman, E., & Nicholas, D. (2019). Scholarly reputation building in the digital age: An activity-specific approach. Review article. *El profesional de la información, 28*(1), e280102. https://doi.org/10.3145/epi.2019.ene.02

Hyland, K. (2001a). Humble servants of the discipline? Self-mention in research articles. *English for Specific Purposes, 20*, 207–226.

Hyland, K. (2001b). Bringing in the reader: Addressee features in academic articles. *Written Communication, 18*(4), 549–574.

Hyland, K. (2002). Authority and invisibility: Authorial identity in academic writing. *Journal of Pragmatics*, 1091–1112.

Knorr-Cetina, K. (1999). *Epistemic cultures: How the sciences make knowledge*. Harvard University Press.

Kress, G., Jewitt, C., Ogborn, J., & Charalampos, T. (2001). *Multimodal teaching and learning: The rhetorics of the science classroom*. Continuum.

Lave, J., & Wenger, E. (1991). *Situated learning: Legitimate peripheral participation*. Cambridge University Press.

León, B., & Bourk, M. (Eds.). (2018). *Communicating science and technology through online video*. Taylor & Francis.

Luzón, M.-J. (2018). Constructing academic identities online: Identity performance in research group blogs written by multilingual scholars. *Journal of English for Academic Purposes, 33*, 24–39.

Luzón, M.-J. (2019). Bridging the gap between experts and publics: The role of multimodality in disseminating research in online videos. *Ibérica, 37*, 167–192.

Luzón, M.-J., & Pérez-Llantada, C. (2022). *Digital genres in academic knowledge production and communication: Perspectives and practices*. Multilingual Matters.

Marwick, A. E., & Boyd, D. (2011). I tweet honestly, I tweet passionately: Twitter users, context collapse, and the imagined audience. *New Media & Society, 13*(1), 114e133.

McNeill, D. (1992). *Hand and mind: What gestures reveal about thought*. Chicago University Press.

Muñoz Morcillo, J., Czurda, K., & Robertson-von Trotha, C. Y. (2016). Typologies of the popular science web video. *Journal of Science Communication, 15*(4), 1–32.

Rowley-Jolivet, E., & Carter-Thomas, S. (2005). Scientific conference Englishes: Epistemic and language community variations. In G. Cortese & A. Duszak (Eds.), *Identity, community, discourse: English in intercultural settings* (pp. 295–320). Peter Lang.

Rowley-Jolivet, E., & Carter-Thomas, S. (2019). Scholarly soundbites: Audiovisual innovations in digital science and their implications for genre evolution. In M.-J. Luzón & C. Pérez-Llantada (Eds.), *Science communication on the Internet. Old genres meet new genres* (pp. 81–106).

Rowley-Jolivet, E., & Carter-Thomas, S. (2020). "Three Minute Theses", ou 3 minutes pour convaincre. Analyse rhétorique d'un nouveau genre universitaire. In F. Domenec & C. Resche (Eds.), *Stratégies et techniques rhétoriques dans les discours spécialisés* (pp. 11–35). Peter Lang.

Shepherd, M., & Watters, C. (1998). The evolution of cybergenres. In *Proceedings of the Thirty-First Hawaii International Conference on System Sciences* (Vol. 2, pp. 97–109). IEEE.

Swales, J. M. (1990). *Genre analysis*. Cambridge University Press.

Thoms, L., & Thelwall, M. (2005). Academic home pages: Reconstruction of the self. *First Monday*.

Toulmin, S. (1958). *The uses of argument*. Cambridge University Press.

Wachsmuth, I., Lenzen, M., & Knoblich, G. (Eds.) (2008). *Embodied communication in humans and machines* (Intro., pp. 1–28). Oxford University Press.

Xia, S. A., & Hafner, C. A. (2021). Engaging the online audience in the digital era: A multimodal analysis of engagement strategies in TED talk videos. *Ibérica, 42*, 33–58.

Part V

Scientific Digital Communication for Research Dissemination: What Lies Ahead?

13

Challenges and Future Directions in Digitally Mediated Research Publication and Dissemination

Vijay K. Bhatia

1 Research Publication in Digital Mode

Digital modes of communication, including social media, have significantly accelerated the speed of publication and dissemination of scientific research. Digital platforms such as blogs, online abstracts and short videos are increasingly being exploited to enhance international visibility. These developments have thus brought into focus the importance of new forms of making research available to larger audiences, use of varied digital modes of communication and, perhaps most importantly, proliferation of digital platforms to publish non-validated preprints online without any peer reviews for immediate visibility within and often beyond not only the community of relevant and intended researchers but also much wider audiences including non-specialist readers. We have witnessed this trend during the period of COVID-19 pandemic, where scientists have raised, argued, discussed and attempted to resolve medical

V. K. Bhatia (✉)
Chinese University of Hong Kong, Hong Kong, People's Republic of China

R. Plo-Alastrué, I. Corona (eds.), *Digital Scientific Communication*,
https://doi.org/10.1007/978-3-031-38207-9_13

and health issues with the kind of pace unheard of in the recent history of mankind. It is true that digital media has brought new opportunities for research publication and dissemination, but at the same time, it has presented unprecedented challenges to academic publishing, especially in the context of what is popularly known as 'open science', which inspires large-scale dissemination of research-based scientific conclusions, sometimes even those that are pre-published scientific observations and findings indiscriminately to specialists as well as non-specialists. On the one hand, it seems to offer more opportunities to disseminate research findings timely, and, on the other hand, it offers a challenge to non-specialists if they need to take such findings as factually established in the scientific community. In this concluding chapter, I would like to highlight (a) some of the key issues in digital communication relevant to research publication and dissemination, (b) identify and discuss some of the challenges and (c) the importance of collaboration in research in international contexts. Finally, I will conclude by suggesting some of the important takeaways as future directions.

2 Digitally Mediated Research Publication and Dissemination: Key Issues

The spread of COVID-19 in the last few years has brought into focus several critical issues in digitally mediated publication and dissemination of scientific research. In what follows, I would like to highlight some of them particularly related to scientific publications (especially in medical and healthcare) on digitally mediated online platforms.

2.1 Speedy Publication and Dissemination

During the early months of COVID-19 pandemic, there was, and it continues to be, an unprecedented surge in the submission of COVID-19–related manuscripts to some of the reputed journals. Bagdasarian et al. (2020) reported that the *New England Journal of Medicine* received up to 40 COVID-19–related publications per day (Jarvis, 2020), whereas the

Journal of the American Medical Association was quoted as receiving up to 235 COVID-19–related manuscripts in a single day. Similarly, *Nature* found approximately 900 COVID-19–related papers, preprints and preliminary reports published prior to 12 March 2020. They pointed out that recognising the need for rapid information on COVID-19, established journals published research findings quickly through expedited review and made them widely accessible. Although they often encourage preprints to make relevant data available to audiences expeditiously, in practice there has always been a marked reluctance to do so. Researchers seemed to be more keen to get speedy publication and immediate dissemination of their findings however tentative they might be, as research dissemination as a key contributor to international visibility has been viewed as one of the most important and valuable concerns of academic life, which is essentially measured in terms of citations, h-index, impact factor, citation analysis and other metrics, such as journal citation reports and several other tools. A good h-index is viewed as the academic footprint of a scholar. If we look at the average time for publishing research digitally online and compare that with publishing in peer-reviewed journals, we find the comparison quite revealing. Peer-reviewed research generally takes 6 to 8 months for publication and dissemination to the relevant academic community for comments and reactions, whereas in digital format, one can get research published almost immediately for a similar outcome. Figure 13.1 sums up the two processes.

It is obvious that the explosion of digital media has made available attractive alternatives to the publication of research findings in journals, particularly in the early months of COVID-19 pandemic when treatment options were required even before the diagnosis of the precise nature of infection. The whole world seemed to be visibly panic-stricken, and, as a consequence, the scientific community succumbed to the temptation of digital alternatives for speedy dissemination of observations and research conclusions. However, digital media also brought with it a negative side as well. Barroga and Matanguihan (2020) in this context rightly claim that the era of COVID-19 has caused distinct shifts in medical and healthcare research publication and dissemination practices, particularly in reframing research design, peer-review practices, adapting and changing research methodologies, managing data

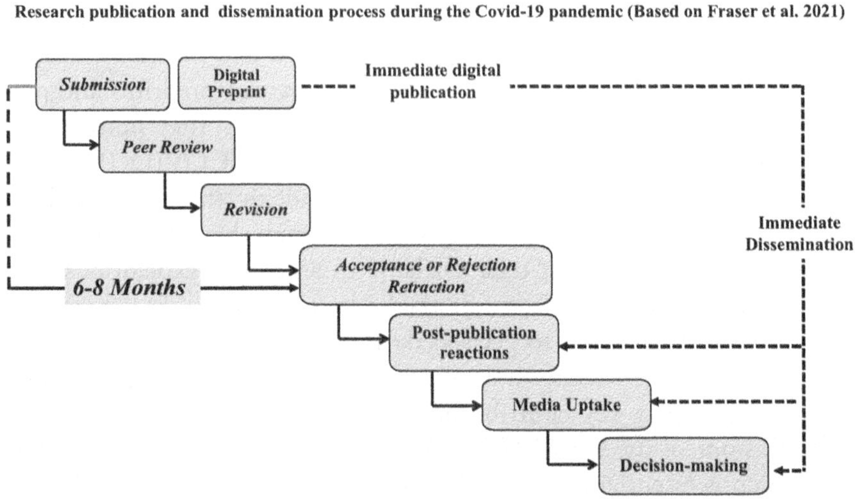

Fig. 13.1 Research publication and dissemination process

collection and standardisation constraints, undermining research standards and ethics, maintaining informativeness and social value and providing guarded peer-review flexibility. Referring to preprints, London and Kimmelman (2020) point out that prompted by the need to disseminate data rapidly, preprints involve less robust interpretation and critical evaluation of scientific reports and many preprints often disregard research reporting standards. They further claim that this pandemic-induced rush for data has resulted in numerous preprint articles published online in digital servers with insufficient scientifically robust data or with trials that do not meet established clinical standards, often without any peer review.

Flanagin (2020) rightly argues that a preprint is posted on a digital platform before peer review to enable authors to seek feedback and comments on research before submitting it to a peer-reviewed journal. Although preprints are required to be revised or updated based on reactions from fellow academics, they often remain on the digital platforms even when they are subsequently published in a journal. Besides, they can be and often are cited and indexed just like reviewed articles, and are likely to be discussed in the news and social media, thus contributing to a much wider dissemination to fellow academics. Such speculative and preliminary studies with sometimes unsustainable results claiming efficient treatments can lead to misleading applications. Preprints are useful for discussion and review, but not for extended dissemination on social media.

In addition to preprint digital platforms to give opportunities for speedy dissemination, we also find a sudden proliferation of predatory journals mushrooming in every discipline. Predatory publishing, as Bocanegra-Valle (2023) convincingly points out, often manages to convince authors to submit their work, often ending up attracting and promoting mediocre research. She also critically assesses questionable editorial practices that may flout mainstream academic publication standards thus undermining the quality of scientific research and the advantages of publishing open access. Similarly, Dinis-Oliveira (2020) confirms that there always is a temptation to compromise on research ethics for speedy publication and academic visibility, which brings us to the second key issue, namely the use of ethically questionable research practices.

2.2 Ethically Questionable Research Practices

Besides infecting millions of people around the world, COVID-19 pandemic seems to have also contaminated the ethical side of academic values particularly in publishing and dissemination of research findings, especially noticeable in medical and healthcare. Dinis-Oliveira (2020) identifies 'Pandemic vs. Paperdemic' as one of the most significant challenges during and even after COVID-19 pandemic that has rocked the world. Researchers have been tempted to sacrifice the scientific rigour for speed, given the urgency to develop new treatments. Jarvis (2020) also claims that science has an ugly, complicated dark side, and the coronavirus is bringing it out. There seems to be a distinct shift in the publication of research output and dissemination practices seen in the online publications such as open science models, where proliferation of preprint servers has made immediate research publication and a much wider accessibility through social media. Digitalisation has changed all aspects of research publication and, more importantly, its dissemination and public engagement. The volume of research papers has grown significantly in the last few years. To make matters more challenging, we have also seen an unprecedented explosion of predatory online journals with little quality control. Even now when the pandemic seems to be under control, this trend has continued giving access to poor and moderate quality research. It seems to have become an accepted norm.

One of the negative aspects of this trend seems to be a low-key acceptance of what is known as ethically questionable practices in research design, as well as its publication and dissemination. The U.S. White House Office of Science and Technology Policy (2000) quite categorically censures such undesirable research practices, which it refers to as 'fabrication', 'falsification' and 'plagiarism' in proposing, performing, reviewing or reporting research results. *Fabrication* refers to making up data or results and recording or reporting them, whereas *falsification* means manipulating research materials, equipment or processes such that the research is not accurately represented. *Plagiarism* is viewed as the appropriation of other's ideas, results or words without giving appropriate credit. For our purpose here, we focus on various forms of fabrication and falsification, both of which have been extensively exploited in COVID-19–related preprints, so much so that one of the editors of the *British Medical Journal*, Richard Smith (July 5, 2021), remarked that "Health professionals and journal editors reading the results of a clinical trial assume that the trial happened and that the results were honestly reported … But oftentimes they may be wrong." He categorically declares that "Time has come to assume that health research is fraudulent until proven otherwise." Gopalakrishna and his colleagues (2022) did an interesting study at Dutch Institutes on engagement of researchers in questionable research practices, and the findings are quite revealing (Fig. 13.2).

It is interesting to note that in their study, more than 50% respondents accepted that they engaged in at least one questionable research practice. There has also been an unprecedented increase in the number of publications in recent years highlighting questionable ethics in medical and healthcare research. An alarming number of studies highlighting some of these concerns have also increased significantly. The titles of many of such

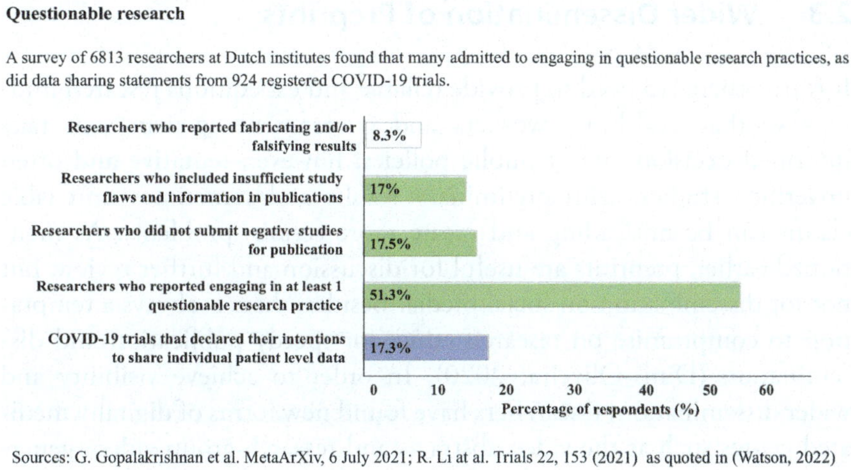

Fig. 13.2 Questionable research

studies underlining this worrying trend in the last one year or so reveal the seriousness of these concerns about such questionable practices. Some of the expressions in these titles reflect the alarming nature of developments related to ethics in publishing and disseminating scientific papers. They include, 'Trust in science', 'Fundamental Shifts in Research Ethics and Peer Review', 'Fraud and Misconduct in Publishing Medical Research', 'The Carnage of Substandard Research', 'Pandemic versus Paperdemic', 'Integrity, Values and Risks of Speed Science', 'Rapid Publications Risk the Integrity of Science', 'How the Pandemic Changed Editorial Peer Review', 'Flawed Preprint Highlights Challenges of COVID Drug Studies', 'An Alarming Retraction Rate for Scientific Publications on COVID-19', 'A Deadly Infodemic: Social Media and the COVID-19 Misinformation', 'Social Media and COVID-19 Misinformation', 'COVID-19-Related Social Media Fake News' and many more. London and Kimmelman (2020) argue that considering the pandemic-related competition for speedy publications has resulted in an increasing number of published articles with insufficient data, or trials, one can perceive weakening of robust interpretation and critical evaluation, which is fundamental to scientific research.

2.3 Wider Dissemination of Preprints

It is true there is a need to provide reliable and expeditious research findings so that healthcare workers and government agencies can take informed decisions about public policies; however, tentative and often unverified studies, with preliminary results and sometimes unreliable claims can be misleading and create more serious problems. As mentioned earlier, preprints are useful for discussion and further review, but not for dissemination on social media. Besides, there is always a temptation to compromise on research ethics for speedy publication and dissemination (Dinis-Oliveira, 2020). In order to achieve visibility and wider dissemination, researchers have found new forms of digitally mediated genres such as the video abstracts and research group videos generally addressed to non-specialist audiences (Dontcheva-Navratilova, 2023; Rowley-Jolivet & Carter-Thomas, 2023), in addition to producing popular versions of scientific research for lay audiences (Benelhadj, 2023, this volume), not to mention scientific blogs to construct researchers' identity (Diani & Freddi, 2023). There have been many instances where research standards have been compromised for speedy publication (Piller & Servick, 2020). An interesting case in point was the retracted preprint, which reported that ivermectin dramatically reduces mortality in COVID-19 patients. Dr Robert Malone even claimed that half-a-million excess deaths were caused by governments that suppressed the early treatment of COVID-19 with drugs including ivermectin and hydroxychloroquine (Rogan, 2022); However, the *Journal of the American Medical Association* (JAMA) at the same time declared that anyone promoting ivermectin for the treatment of COVID-19 should be disciplined by state medical boards (Rubin, 2022). This case highlights the differences in interpretations by relevant scientific community members and the risks associated with such tentative preprint publications.

Watson (2022, p. 2), in a similar context, points out that "A staggering 19,389 articles about Covid-19 were shared in the first four months of the pandemic, a third of which were preprints, unvetted and unfiltered for all to see." She also refers to the statement made by Maria van Kerkhove (COVID-19 technical lead at the World Health Organisation (WHO))

who claimed that "she and her teams would weigh up 'every shred of evidence,' forming a position after critiquing the fine print of each study on a topic, and never looking at one paper alone." Van Kerkhove admits that overall preprints are "accelerating the pace of research and directly informing public-health policies, but they can misdirect a policy or lead you astray if you don't stay rooted in the totality of the science." A number of questionable preprints that turned out to be fraudulent, yet continued to reverberate long after they were either withdrawn or refuted, illustrates the danger of preprints that never pass peer review.

However, Watson (2022, p. 2) further points out that "despite the drawbacks and deadly consequences, there is little doubt that preprint publishing is here to stay. The question is how science will handle it." She refers to Sheldrick, a medical practitioner and data sleuth at the University of New South Wales, Australia, who comments, "We are down a pathway of open science, and that pathway is going to accelerate... Our choice is not whether it occurs or not; our choice is whether it occurs responsibly." In order to discuss the issues related to responsibility in open science, it is in order to first briefly mention the role of social media and its implications, particularly in the context of preprints.

2.4 Role of Social Media

Social media, such as Facebook, YouTube, WhatsApp, Instagram, Twitter, Tik-Tok and several other platforms have billions of users in the world, and most of them have become a significant communication tool not only for information generation and dissemination, but more critically for consumption of information whether false or fabricated without any scrutiny. Ordinary readers active on social media tend to not only passively consume disinformation or misinformation but also disseminate and even generate such fabrications. Monensen (2020) rightly argues that bad science is like a virus that starts small, but if it is shared extensively, it can cause global disruption. COVID-19 pandemic was perhaps one of the most critical periods in the recent history of open science for the spread of web-mediated disinformation and misinformation about science. Both social media companies and other digital platform users played a substantial role in the dissemination of popular versions of scientific research without any checks for accuracy, thus contributing to the creation of what the WHO called 'infodemic' for "too much information or false and misleading information" that "causes confusion, risk taking behaviours … and mistrust of health officials." It has also been claimed that social media was partly responsible for increasing vaccine hesitancy, lowered vaccination rates and for causing preventable deaths (de Albuquerque et al., 2021; Razai et al., 2021). Social media companies beyond any doubt are partly responsible for accelerating the COVID-19 infodemic. The stakes in this as we know are very high, and failure to manage health misinformation and disinformation should be viewed as a public health crisis. Mainstream media coverage also seem to have added to the problem, as most major news agencies, with reporters and editors without much or limited medical knowledge or public health training, have been required to cover the pandemic and manage scientific terms, methodologies and research, and then identify, evaluate and review credibility of scientific resources and information, which often adds further confusion to much of preliminary scientific observations and tentative findings (Pazzanese, 2020).

3 Growing Awareness of Collaborative Research and Participatory Culture

Having identified and discussed the negative side effects of digitally mediated communication, let us look at some of the positives as well. Digital communication, particularly in international contexts, has inspired an increasing number of initiatives in collaborative research and participatory interdisciplinary culture. Let us briefly examine the processes of and implications for the future of collaborative research not only in science and technology but also in many other domains of research, including social sciences and humanities. It is obvious that much of research-related communication, especially in international contexts, is primarily taking place in digital environments. Even most of the established journals and publishers are using online platforms, with emphasis on international collaboration in a participatory environment. Drawing on my experience of international collaborations across more than twenty countries involving colleagues from a range of disciplinary backgrounds (Bhatia, 2017; Bhatia et al., 2012), I would like to illustrate how collaborative research offers a valuable resource that not only contributes to academic visibility but, at the same time, provides opportunities to develop multiperspective and multidimensional frameworks often leading to a more comprehensive and insightful understanding of discursive and professional practices particularly in the context of the present-day digitally mediated global world, which enables speedy communication for consultation and dissemination of research insights, in addition to encouraging collaborative authorship and multiple opportunities for publication. Let me identify and briefly discuss some of the key aspects of this kind of collaboration in a participatory culture.

One of the most important factors that makes collaboration successful is the use of a multiperspective approach, specifically useful for an inter- and multi-disciplinary perspective integrating varying frameworks, methods and expertise. It is crucial to remember that the loyalty of the individual researchers in the team often is to the data being used and analysed for research and not necessarily to any specific framework. A successful initiative crucially depends on accommodating particular interests, divergent, sometimes irreconcilable mindsets, conflicting ideologies, distinct identities, diversity of disciplinary cultures and ways of doing things independently, uniquely and distinctly. With digital technology it is also possible for individual members of the team to exercise their freedom to follow one's own convictions within the broad constraints of the overall objectives of the collaboration. Successful and sustained collaboration thus rests on the awareness of participatory culture on the part of individual researchers, and digital forms of communication make it swift and smooth. Collaboration invariably inspires co-authorship and speedy communication, which enhances the visibility and dissemination of research outcomes. In my view, collaborative research contributes significantly to international visibility, particularly in the context of present-day digitalisation of academic and research communication. Co-authorship patterns in bibliometric databases and the other networks distinctly indicate a high percentage of multi-authored publications, not only in science and technology, but increasingly in social and human sciences too.

4 Future Directions

In the preceding sections of the chapter, I have tried to highlight some of the challenges that digital forms of publications as open science have brought into focus, some of which include the need to reform the review process to manage questionable ethical practices. Rapid publication through expedited review and online preprints are important resources for the scientific community; however, they also come with the risks of unverified observations and tentative findings, especially when the insights, findings and conclusions are eagerly awaited for applications and policy decisions. In today's context, when digital publications have become popular, even the members of the scientific community and non-specialists, who primarily depend on the social media to get the fruits of science, need to consider many of these publications carefully and consider additional steps to make sure that such expedited publications and preprints are a credible and reliable resource. Engberg (2023), accounting for knowledge communication through digital media, makes an interesting distinction between 'open science' and 'citizen science' and proposes a number related concepts such as 'infotainment', 'dissemination', 'popularisation' and 'pedagogic communication', all of which need to be distinctly identified in the process of interpretation of scientific research.

One needs to adopt a more useful multiperspective view of such observations and findings to make sure the extent to which they are reliable. Chater (2020) rightly argues that we often struggle to organise this confusing mass of partial and confusing information into a coherent interpretation. We make decisions as if our interpretation is true without entertaining alternatives, which can lead to *discursive illusions* (Bhatia, 2015) that often inspire disastrous decision-making. We need to search actively for multiple interpretations. Breeze (2023) addresses this issue discussed on social media, examining how individuals ground their knowledge claims on subjects they know little about and emphasising the role of group affiliations in such discussions.

It is true that open science with much wider dissemination possibilities enhances visibility of researchers, but at the same time it has the responsibility that demands greater transparency of procedures and data. It requires a concerted effort on the part of digital platforms to challenge the weak peer-review systems that allow non-validated research publications. It also needs to ensure that there is a valid mechanism in place to restrict wider circulation of preprints through social media. Also, preprint servers need to do their part to limit access to subscribers with an academic affiliation for the purpose of getting feedback and input from peers and to encourage authors to post an updated version of the manuscript as and when one gets comments and reactions from peers or reviewers and to include feedback from reviewers and editors. It may also be possible to add a notification on each manuscript that has not been peer-reviewed.

On the part of individual scholars, it is necessary to adapt available digital resources to disseminate credible research, and not preliminary non-validated research findings. Dinis-Oliveira (2020) rightly argues that "a balance is needed between the benefits of the rapid access to new scientific data and the threat of causing panic or erroneous clinical decisions based on mistakes or misconduct. The truth is that 'scientific research has changed the world' but now, and more than ever, 'it needs to change itself'. A pandemic with a 'paperdemic' will be even more complicated to manage if it progresses in an uncontrolled manner and is not properly scrutinized." A shocking case in point was the unfortunate misinterpretation of a letter published in *Lancet Respiratory Medicine* in the early days of the pandemic by a group researcher that raised safety concerns about the use of ibuprofen to treat coronavirus symptoms. It was just an opinion, but WHO misinterpreted it as a recommendation. However, they soon realised their mistake but not before it had created considerable confusion in several countries, including France. Maria van Kerkhove (quoted in Watson (2022, p. 2)) points out that it is crucial to weigh every shred of evidence, forming a position after critiquing the fine print of each study on a topic, never looking at one paper alone. Gilbert (2022) interestingly refers to the parable of the six blind men and the elephant, which to me is the most important argument for a multiperspective analysis and interpretation of language in use, including scientific arguments and claims. In the parable, each blind man uses touch to carefully examine part of an elephant. Each of them reaches a correct conclusion about what was observed. None of the blind men are guilty of fraud, dishonesty or misinformation. However, none of the blind men reaches a correct conclusion about the totality of the elephant. In the case of current infodemic, it is the most critical strategy to avoid tunnel vision, sampling bias and the inability of a single person to know everything about a complex problem. The parable seems to be applicable to uncertainty in medical and healthcare interpretation as well. To sum up, open science is firmly established in research publication and dissemination practices, but at the same time it needs to reform or reframe research design processes, peer-review practices, encourage multiperspective interpretations of research findings, inspire better management of data collections processes and procedures and discourage undermining of research standards and ethics.

References

de Albuquerque Veloso Machado, M., Roberts, B., Wong, B. L. H., van Kessel, R., & Mossialos, E. (2021). The relationship between the COVID-19 pandemic and vaccine hesitancy: A scoping review of literature until August. *Frontiers in Public Health, 9*, 747787. https://doi.org/10.3389/fpubh.2021.747787. [Medline: 34650953].

Bagdasarian, N., Cross, G. B., & Fisher, D. (2020). Rapid publications risk the integrity of science in the era of COVID-19. *BMC Medicine, 18*, 192. https://doi.org/10.1186/s12916-020-01650-6

Barroga, E., & Matanguihan, G. J. (2020). Fundamental shifts in research, ethics and peer review in the era of the COVID-19 pandemic. *Journal of Korean Medical Science, 35*(45), e395. https://doi.org/10.3346/jkms.2020.35.e395

Benelhadj, F. (2023). Dissemination of knowledge during the Covid 19 pandemic: A conceptual metaphor analysis of research and popular articles. In R. Plo-Alastrué & I. Corona (Eds.), *Digital scientific communication: Identity and visibility in research dissemination*. (Ch.4). Palgrave.

Bhatia, A. (2015). *Discursive illusions in public discourse: Theory and practice*. Routledge.

Bhatia, V. K. (2017). *Critical genre analysis: Interdiscursive performance in professional practice*. Routledge.

Bhatia, V. K., Candlin, C. N., & Gotti, M. (2012). *Discourse and practice in international commercial arbitration*. Ashgate Publishing Limited.

Bocanegra-Valle, A. (2023). Predatory journals: A potential threat to the dissemination of open access knowledge. In R. Plo-Alastrué & I. Corona (Eds.), *Digital scientific communication: Identity and visibility in research dissemination*. (Ch.6). Palgrave.

Breeze, R. (2023). "Not one of our experts." Knowledge claims and group affiliations in online discussions of the COVID-19 vaccine. In R. Plo-Alastrué & I. Corona (Eds.), *Digital scientific communication: Identity and visibility in research dissemination*. (Ch. 2). Palgrave.

Chater, N. (2020). Facing up to the uncertainties of COVID-19. *Nature Human Behaviour, 4*, 439. https://doi.org/10.1038/s41562-020-0865-2

Diani, G., & Freddi, M. (2023). Authorial stance and identity building in weblogs by law scholars and scientists. In R. Plo-Alastrué & I. Corona (Eds.), *Digital scientific communication: Identity and visibility in research dissemination.* (Ch.5). Palgrave.

Dinis-Oliveira, R. J. (2020). COVID-19 research: Pandemic *versus* "paper-demic", integrity, values and risks of the "speed science". *Forensic Sciences Research, 5*(2), 174–187. https://doi.org/10.1080/20961790.2020.1767754

Dontcheva-Navratilova, O. (2023). Video abstracts for increasing researcher visibility. In R. Plo-Alastrué & I. Corona (Eds.), *Digital scientific communication: Identity and visibility in research dissemination.* (Ch.8). Palgrave.

Engberg, J. (2023). Between infotainment and citizen science: Degrees of intended non-expert participation through knowledge communication. In R. Plo-Alastrué & I. Corona (Eds.), *Digital scientific communication: Identity and visibility in research dissemination.* (Ch.7). Palgrave.

Flanagin, A. (2020). Public access and open access in scientific publication. In S. L. Christiansen, C. Iverson, A. Flanagin, et al. (Eds.), *AMA manual of style: A guide for authors and editors* (11th ed.). Oxford University Press.

Fraser, N., Brierley, L., Dey, G., Polka, J. K., Pálfy, M., Nanni, F., & Coates, J. A. (2021). The evolving role of preprints in the dissemination of COVID-19 research and their impact on the science communication landscape. *PLoS Biology, 19*(4). https://doi.org/10.1371/journal.pbio.3000959

Gilbert, B. (2022). Controversy over ivermectin and COVID-19: Six blind men and an elephant. *The Southwest Respiratory and Critical Care Chronicles, 10*(43), 52–55.

Gopalakrishna, G., Wicherts, J. M., Vink, G., Stoop, I., van den Akker, O. M., ter Riet, G., & Bouter, L. M. (2022). Prevalence of responsible research practices among academics in The Netherlands. *F1000Research 2022, 11*, 471. https://doi.org/10.1371/journal.pone.0263023

Jarvis, C. (2020, March 17). Journals, peer reviewers cope with surge in COVID-19 publications. *The Scientist.*

London, A. J., & Kimmelman, J. (2020, April 23). Against pandemic research exceptionalism. *Science* 368(6490), 476–477. https://doi.org/10.1126/science.abc173

Malone, R. (2022). The Joe Rogan Experience. Episode #1757—Dr. Robert Malone, MD. https://open.spotify.com/episode/3SCsueX2bZdbEzRtKOCEyT

Monensen, J. F. (2020, April 28). Science has an ugly, complicated dark side and the Coronavirus is bringing it out. *The Mother Jones Daily Newsletter.* https://www.motherjones.com/politics/2020/04/coronavirus-science-rush-to-publish-retractions/

Pazzanese, Ch. (2020, May 8). Battling the 'pandemic of misinformation'. *The Harvard Gazette.*https://news.harvard.edu/gazette/story/2020/05/social-media-used-to-spread-create-covid-19-falsehoods/

Piller C., & Servick, K. (2020, June 4). Two elite medical journals retract coronavirus papers over data integrity questions. *Science.* https://doi.org/10.1126/science.abd1697

Razai, M. S., Osama, T., McKechnie, D. G. J., & Majeed, A. (2021). Covid-19 vaccine hesitancy among ethnic minority groups. *BMJ (Clinical Research Ed.), 372,* n513. https://doi.org/10.1136/bmj.n513

Rowley-Jolivet, E., & Carter-Thomas, S. (2023). Research visibility and author ethos: A comparative study of 3MT presentations and research group videos. In R. Plo-Alastrué & I. Corona (Eds.), *Digital scientific communication: Identity and visibility in research dissemination.* (Ch.12). Palgrave.

Rubin, R. (2022). When physicians spread unscientific information about COVID-19. *JAMA, 327*(10), 904–906.

Smith, R. (2021, July 5). Time to assume that health research is fraudulent until proven otherwise. *The BMJ Opinion.*

The U.S. White House Office of Science and Technology Policy. (2000). *Definition of research misconduct.* The Office of Research Integrity. https://ori.hhs.gov/definition-research-misconduct

Watson, C. (2022). Rise of the preprint: How rapid data sharing during COVID-19 has changed science forever. *Nature Medicine, 28,* 2–5.

Index[1]

[1] Note: Page numbers followed by 'n' refer to notes.

R. Plo-Alastrué, I. Corona (eds.), *Digital Scientific Communication*,
https://doi.org/10.1007/978-3-031-38207-9

SPRINGER NATURE

GPSR Compliance

The European Union's (EU) General Product Safety Regulation (GPSR) is a set of rules that requires consumer products to be safe and our obligations to ensure this.

If you have any concerns about our products, you can contact us on ProductSafety@springernature.com

In case Publisher is established outside the EU, the EU authorized representative is:

Springer Nature Customer Service Center GmbH
Europaplatz 3
69115 Heidelberg, Germany

The manufacturer's authorised representative in the EU is Springer Nature Customer Service Centre GmbH, Europaplatz 3, 69115 Heidelberg, Germany. If you have any concerns regarding our products, please contact ProductSafety@springernature.com

Printed and bound by CPI Group (UK) Ltd, Croydon, CR0 4YY
28/09/2026
02224074-0009